Engineering and Technology
In
Ancient India

by

Prof. Ravi Prakash Arya

Chair Professor
Maharshi Dayanand Saraswati Chair) UGC)
Maharshi Dayanand University, Rohtak

AMAZON BOOKS, USA

In association with

INDIAN FOUNDATION FOR VEDIC SCIENCE

H.O.1051, Sector-1, Rohtak, Haryana, India
Delhi Contact Ph. Nos.: 09313033917; 09650183260
Emails: vedicscience@gmail.com; edicscience@rediffmail.com
web: https://vedic-sciences.com

First Edition

Kali era 5123 (c. 2022)
Kalpa era 1,97,29,49,123
Brahma era 15,55,21,97,29,49,123

ISBN : 9788194759300

© Author

Printed by

Indian Foundation for Vedic Science, 1051, Sector-1, Rohtak-124001, Haryana

Contents

Preface

India is credited with a highly advanced knowledge system since ancient times. Vedas and post-Vedic literature composed in Sanskrit has been the storehouse of various branches of knowledge on science, spirituality, philosophy, culture, civilisation, economy, polity, law and management, ayurveda, yoga, and pharmacy. This knowledge was documented from time to time and carried forward for future generations until the mass-scale destruction of Sanskrit manuscripts to plunder and loot and devastation of Nalanda, Takshshila, Vikramshila and Ujjain libraries. Still, a huge number of Sanskrit manuscripts on various subjects are found preserved in various museums and libraries in India and abroad. Nearly 1.5 million manuscripts are held in the libraries and museums of America and Europe, and more than 5 million are preserved in India.

Thus, it can be maintained without an iota of doubt that India contributed immensely towards the origin and evolution of various disciplines of the modern age. However, it is a matter of great concern that India's contribution to various fields of learned studies is often neglected, underestimated, negated, derecognised, distorted and twisted to such an extent that the modern generation of Indians had a tough time believing that Indians could contribute to various fields of learned studies. However, they could produce literature on speculative metaphysics, mysticism and art. Thus, much confusion persists under the Eurocentric vision of natural and social sciences.

On the other hand, impartial studies and research carried out without any prejudice or preconceived notions prove that India had a long-glorified intellectual tradition. Just imagine the modern world without the Vedic mathematical revolution of ancient India. The fact is that the scientific process during European Renaissance was a result of this mathematical revolution that had its origin in India. Modern thinkers and scientists like Goethe, Emerson, Thoreau, Jung, Oppenheimer, Herder, and Schrodinger, to name a few, have acknowledged their debt to ancient Indian achievements in science, technology and philosophy. Accounts of ancient scholars, thinkers and historians like Aristotle, Arrian, Megasthenes, Clement of Alexandria, and Apollonius of Tyana among the Greeks; Andalusi, Al-Beruni, Al-Uqlidisi among Islamic scholars, Fa-Hien, Hiuen Tsang, and I-tsing among Chinese,: Leonardo Fibbonacci, Pope Sylvester II, Roger Bacon, Voltaire, and Copernicus from Europe prove the existence of civilisational knowledge system/intellectual tradition in ancient India.

Even though Indian culture has a long and glorified intellectual tradition, the education system in India fails Indians to know about their roots, traditions, culture and sciences. The science and social science courses in India taught us about Democritus, Archimedes, and Newton, Plato, to name a few, but nothing about Manu, Baudhāyana, Kaṇāda, Āryabhaṭṭa Bhāskara, and Brahmagupta, Chāṇakya and others. Also, non-academic literature on this glorious Indian intellectual tradition is missing. It is because science and other texts in the

Indian education system are primarily Eurocentric. Several historians have recognised these omissions of scholarly contributions of non-western cultures in various fields of learned studies. For example, Joseph Needham, a historian of science from the UK, recognised these omissions as deeply unjust to other civilisations. He says, 'Unjust here means both untrue and unfriendly, two cardinal sins mankind cannot commit with impunity.'

So far as the Indian scenario is concerned, there is a dearth of scholarly, systematic and comprehensive study on intellectual tradition in Ancient India, except a few fragmentary references and studies. This has greatly disappointed scholars who want to research the intellectual tradition of ancient India. The echo of this disappointment is clearly heard in the words of James E. McClellan III and Harold Dorn, Science and Technology in World History: An Introduction,

The Johns Hopkins University Press, 1999. He expresses his concern as follows:

'In recent decades, the scholarly study of science and civilisation in China has influenced historians concerned with the history of science and technology in India. But, alas, no comprehensive synthesis has yet appeared to match the studies of China" "Given its technological complexity, India actually underwent an astonishing process of deindustrialisation with the coming of formal British rule in the nineteenth century'.

A similar concern is expressed by another Dutch scholar Claude Alvares from Holland. Nonetheless, he speaks in the exact words of James E. McClellan III and Harold Dorn.

'While there was certainly no dearth of historical material and scholarly books as far as Chinese science and technology were concerned—largely due to the work of Dr Joseph Needham, reflected in his multi-volumed Science and Civilisation in China—in contrast, scholarly work on Indian science and technology seemed to be almost non-existent. What was available seemed rudimentary, poor, unimaginative, wooden, more filled with philosophy and legend than fact.'

From the foregoing, it is amply clear that the intellectual tradition of India has been subjected to utter neglect and stagnation. So Indian accounts are absent in the academic texts of India. As such, there is an ardent need to initiate a process to unlock these systems of knowledge that are currently atrophied due to centuries of neglect and stagnation. Given the above, the present study was undertaken. This work is a humble attempt to apprise the readers about the advancement of Engineering and Technology in Ancient India. I should not fail here to acknowledge the assistance rendered by Mr. Alois Heinrich from Stuttgart, Germany, in proofreading and giving several precious suggestions for accomplishing this uphill task. I hope the students and scholars find this book equally exciting and helpful in their academic pursuits in Ancient Indian Civilizational Knowledge System.

1

Introduction

Science

Science found a prominent place in ancient India. We came across such references in the Mahabharata (Śānti Parva, 205.3), where science was highly significant. Pointing out the great importance of science, Vyāsa says:

प्रज्ञया मानसं दुःखं हन्याच्छरीरमौषधैः ।
एतद् विज्ञानसामर्थ्यं न बालैः समतामियात् ॥

prajñayā mānasaṁ duḥkhaṁ hanyāchchharīramauṣadhaiḥ
ētad vijñānasāmarthyaṁ na bālaiḥ samatāmiyāt ॥

It is the significance of science that we can cure psychological problems through psychotherapies and physical diseases through the administration of medicines, nor would we have wept like children when faced with these diseases.

Stressing upon the importance of knowledge and its realization or practical applicability (science), Vyāsa (Śānti Parva, 326.22) observes that without the Jñāna (knowledge) and Vijñāna (science), it is challenging to attain Mokṣa.

न बिना ज्ञानविज्ञाने मोक्षस्याधिगमो भवेत् ।

na binā jñānavijñāne mokṣasyādhigamo bhavet ।

In Anuśāsana Parva of the Mahābhārata (8.8), Bhiṣma Piāmaha while narrating the qualities of scholars, gives utmost importance to those scholars who are well versed in science.

ये चापि तेषां श्रोतारः सदा सदसि सम्मताः ।
विज्ञानगुण समुपास्तेभ्यश्च स्पृहयाम्यहम् ॥

ye chāpi teṣāṁ śrotāraḥ sadā sadasi sammatāḥ /
vijñānaguṇa samupāstebhyaścha spṛhayāmyaham //

Engineering

The contents of the Vedas have been divided into four categories: Jñāna Kāṇḍa, Vijñāna Kāṇḍa, Karma Kāṇḍa and Upāsanā Kāṇḍa.

1. **Jñāna Kāṇḍa** deals with the theory aspect of the information. Ādi Saṅkarācārya (1168 BC-1136 BC) defines and differentiates into Jñāna and Vijñāna as under:

ज्ञानं विषयः विषयानुभूतिर्विज्ञानम् । विज्ञानमानन्दं ब्रह्म ।

jñānaṁ viṣayaḥ viṣayānubhūtirvijñānam /
vijñānamānandaṁ brahma /

That is, jñāna is bare information or knowledge of an object or a thing, but science is experiential knowledge or the realisation of that particular thing or an object. Experience or realization of Brahman is blissful.

2. **Vijñāna Kāṇḍa** deals with the realizational, experiential or practical aspect of the information. Should the Jñāna be a theory, its practice is Vijñāna. Suppose the Jñāna is associated with the spiritual element. In that case, the realization will be treated as its Vjñāna and should it be related to the material aspect, its practical knowledge will be called as Vjñāna.

The difference between Jñāna and Vijñāna can be understood by the statement of Ādi Saṅkarācārya (1168 BC-1136 BC) cited above.

Maharishi Dayananda Saraswati (1824 AD-1883 AD),

one of the leading Vedic scholars and social reformers of 19th century India, also defines Vjñāna as:

तस्य परमेश्वरादारभ्य तृणपर्यन्तं पदार्थेषु साक्षात्बोधान्वयत्वात् ।

tasya parameśvarādārabhya tṛṇaparyantaṁ padārtheṣu sākṣātbodhānvayatvāt /

The domain of science entails the experience or realisation of everything from the root of grass to God.

Thus, we see that science was defined in India in terms of realizational, experiential and experimental knowledge. Modern science defines it as observation-based science. Whatever is observed through experiments in the lab is called science. Realization or experience is a more profound concept than observation. Observational power is limited. Everything cannot be observed, e.g. observation of abstract things is impossible. Observation of hunger and thrust is not possible. They can be experienced. The experience of a Yogī or mystic can be verified by another mystic or a yogī. It cannot be observed by doing experiments in the lab. The realization of even those things which cannot be observed is possible. Observation is a physical concept, whereas realisation is a spiritual concept. The realisation is oneness with the thing or an object realised. When a Yogī realises Brahman, he becomes virtually Brahman.

यो ब्रह्म जानाति स ब्रह्मैव भवति ।

yo brahma jānāti sa brahmaiva bhavati /

One who knows the Brahman becomes virtually Brahman.

3. **Upāsanā Kāṇḍa** indicates the dedication of the seeker to realize or practically apply the information known to him.

4. **Karma Kāṇḍa** deals with the efforts to modify or develop something based on the information known to the individual. Thus, according to the Vedic philosophy, engineering is an effort for the Sanskāra (modification) of dravyas (matter and soul) into devising/making something, and technology is the result of those efforts for Sanskāras. Mimānsā philosopher, Jaimini (3100 BC), defines engineering and technology as under:

द्रव्यसंस्कार कर्मसु परार्थत्वात् फलश्रुतिरर्थवादः स्यात् ॥

dravyasaṁskāra   karmasu   parārthatvāt phalaśrutirarthavādaḥ syāt // 4.3.1

Engineering is the sanskāra of the matter (made of five gross elements) for developing technology to benefit humanity so that the technology may provide desired results.

Further defining the purpose of engineering, Jaimini says:

द्रव्याणां तु क्रियार्थानां संस्कारः क्रतुधर्मः स्यात् ॥

dravyāṇāṁ tu kriyārthānāṁ saṁskāraḥ kratudharmaḥ syāt // 4.3.8

The main objective of the क्रतुधर्मः (engineering) is the sanskāra (processing) of the material things for developing various technologies.

Here it may be known that the modern technique of refinement, modification, purification, and processing the material things is known in Vedic science as *sanskāra*. The main thrust of modern science has been the *sanskāra* of material things and the development of technology to promote the convenience and comfort of human beings. Here there is a fundamental difference between Vedic Science and modern science. Vedic science primarily stresses the *sanskāra* of human beings before going for the sanskāra of material things. According to Vedic science, a human being who has not gone through the certain process of *sanskāras* remains in his crude form and is recognized as no better than an animal. As such, a human being is not considered an actual social being. Just as material things cannot be used efficiently so long as they are in their crude form, similarly, a human being, without undergoing the process of Sanskāras, cannot become a social, cultural, and civilised being.

As for the practical application, the material things are subjected to undergo a particular process of refinement or *sanskāra*, like oil in its crude form cannot be used. Still, after undergoing the process of refinement, it becomes usable. Thus, converting oil from its crude form to a refined format is known as engineering. Similarly, the processing of other material things for the use of human beings is also known as engineering. Modern science has ignored the need for *sanskāra* of humans in the race for developing technology. That is why, today, in the age of machinery, human beings are also considered as good as machines. They are equated with lifeless things, so there is always a talk of human resource development and never human development. The concept of human resource development is embedded in the very philosophy that gives human beings no better value than other natural resources, which are often exploited for the benefit of the exploiter, and so are human beings. However, there is a fundamental difference between the exploitation of natural sources and human resources. The exploitation of natural resources takes place uniformly for the benefit of humanity at large. But so far as human beings are concerned, the poor and weak are exploited by the powerful and wealthy.

So, Vedas are equally concerned about the *sanskāra* of human beings and material things for developing technologies beneficial to humankind. Vedic seers never considered human beings as resources or means but as ends. Everything is directed towards the development and elevation of human beings.

Moreover, Vedic seers have a humanitarian and ethical approach to everything. Their main aim was to relieve human beings permanently of the suffering, social injustice, deprivation, and all such problems that haunt them.

Modern Science has developed many aids for the comfort of human beings, yet it has a limited approach. Its approach is not Universal. Modern science has no moral, ethical, or humanitarian foundation. It is founded on the laws of survival of the fittest in the struggle that ensues among human beings on the globe. Modern science wants to define and describe everything in the light of struggle. It has no remedy to put an end to this struggle. Contrary to this, the main aim of Vedic science is to put an end to the struggle among human beings, between human beings and animals and human beings and ecology or environment and prepare a stage for peace, friendship, fraternity and co-existence of every being on the Earth in harmony with nature. Vedas give a clarion call for the principles of mutual understanding and cooperation among human beings.

स नौ भुनक्तु सह वीर्यं करवावहै ।
तेजस्विनावधितमस्तु मा विद्विषावहै ॥

sa nau bhunaktu saha vīryaṁ karavāvahai |
tejasvināvadhitamastu mā vidviṣāvahai ||

"Let there be cooperation in eating and gaining power. Let each one of us become illuminated with knowledge. Let us not envy or struggle with each other."

Since modern science has its foundation in the laws of struggle, its talk about exploiting natural and human resources aims to comfort those who can survive the global struggle. We can see that despite high advancement in the area of technology and the production of each and everything, the benefits are not reaching each human being on the globe due solely to the absence of moral and humanitarian values. We have more than a sufficient amount of food grains piled up in our stocks; still, there is poverty and hunger on the globe. On the one hand, the excess food grains are being disposed of in the sea; on the other hand, human beings and other animate beings are dying of hunger. Modern Science has though blessed modern humanity with great prosperity; still, more than 50% of human beings on the globe are below the poverty line and suffering from hunger, thrust, malnutrition, the phenomenon of coldness and hotness, and a dearth of essential things for their survival. One can say unhesitatingly that modern science has failed in performing the *sanskāra* of human beings. It could develop and devise machines but not good human beings.

Today we see more and more extensions of material technology but less awareness of science; during the Vedic period, understanding of science dominated over technological know-how.

Today yantrika technology is dominant; in the Vedic

period, Mantrika and Tantrika technology was dominant.

Today we are in a blind race for material gains and causing environmental hazards like global warming, depletion of the ozone layer, and environmental pollution. Still, the Vedic people emphasized establishing harmony between nature and human beings.

Today we are emphasizing Sanskara of material things. In the Vedic period, the Sanskara of human beings was highlighted.

In the modern scientific period, advancement is measured/mapped through technology, but in the Vedic period, it was mapped through scientific awareness. After implementing the WTO proposal, Lauren Summer, president of the World Bank, asked America to transfer hazardous industries to developing and under-developed, so those rich countries may not face any health hazards due to leakage of chemicals, etc, and could remain safe. Because life is considered very expensive in developed countries and very cheap in developing countries. People in developing and underdeveloped countries are like gunny pigs for doing experiments in labs. In India also, we find a chain of chemical factories in Gujrat, starting from Vapi to Mehsana. Bhopal gas is one such examples.

2

Nature and Scope of Vedic Science and Technology

The Vedas are the oldest literary records of the significant advance made by humanity on the globe. The sublime and the rich poetry of the Vedas, the antiquity, perfectness, copiousness, exquisite refinedness and wonderfulness of the structure of the language they are composed of all point out to the highly cultured and civilised nature of the race within which this rich, and everlasting heritage was originated. The term civilisation is a pointer to an advanced stage of living, where people can lead a comfortably settled and prosperous life and have spare time for innovative activities in various fields.

Under the circumstances, it is natural for modern generations to evince interest in the social conditions of the highly advanced race that is said to have represented the remotest past of this country. Now, when science and technology have become the way of life, and everything is tested and examined in the light of scientific ideas, it becomes more interesting to know about the nature and type of science the Vedic people professed and the nature and style of the technology they developed and devised.

Origin of science: The origin of science always takes place in the outlooks of individuals or society towards the creation of the world of the living (Cetana) or non-living

(Jaḍa) things. The innovations, inventions, or discoveries in the field of science and technology take their shape or derive inspiration mainly from the philosophical hypothesis formulated by an individual or society towards the origin or composition of the world. It depends entirely upon the philosophical viewpoint of the innovator, inventor or discoverer as to how to philosophise the world around him, whether he takes it as a composition of Prakṛti (matter) only or as an outcome of a mere accident of Puruṣa (self), and Prakṛti or as an outcome of accident of Puruṣa and Prakṛti caused by the will of Almighty Brahman. So, one's range of innovations, inventions, or discoveries is always encompassed by the scope of one's philosophical hypothesis. For instance, in modern times, the living or non-living world is considered the creation/composition of matter. This is why modern science studies matter, the physical properties of which are studied in Physics and Chemical properties in Chemistry. Modern technology is also devised based on the advancement made by material science.

The Vedic Origins of Science: In the Vedic times, science was developed around the philosophical innovations of the Vedic Ṛṣis. According to the philosophical concept of the Vedic Ṛṣis, the world was not the composition of matter only; it was considered to have represented by Prakṛti, Puruṣ and Brahman in various capacities.[1] It evolved on account of the accident of Puruṣa

---

[1]   See *RV.* 1.26.4.

द्वा सुपर्णा सयुजा सखाया समानं वृक्षं परिषस्वजाते
तयोरन्यः पिप्पलं स्वाद्वत्त्यनश्नन्यो अभिचाक्शिति॥
*dvā suparṇā sayujā sakhāyā samānaṁ vṛkṣaṁ pariṣasvajāte
tayoranyaḥ pippalaṁ svādvattyanaśnanyo abhicākśiti.'*

and Prakṛti, the accident being materialised by the will (Saṁkalpa) of the Brahman.[2] The accident of Puruṣa with Prakṛti, which is the name of the equilibrium of Sattva, Rajas and Tamas qualities[3], brings about the disequilibrium in the triad of qualities (i.e. guṇatraya), which ultimately causes the Vikṛti of Prakṛti, or evolution of matter as under:

प्रकृतेर्महान् महतोऽहंकारोऽहंकारात् पञ्चतन्मात्राणि उभयमिन्द्रियं पञ्चतन्मात्रेभ्यः स्थूलभूतानि पुरुष इति पञ्चविंशतिर्गणः ।

Prakṛtermahān mahato'haṁkāro'haṁkārāt pañcatanmātrāṇi ubhayamindriyaṁ pañcatanmātrebhyaḥ sthūlabhūtāni puruṣa iti pañcaviṁśatirgaṇaḥ

That is two beautiful friendly birds (*Ātman* and *Brahman*) are sitting together on a tree (*Prakṛti*). One of them tastes the sweet fruits and another sits indifferently looking at or acts as a mere witness to the former'.

See also *Chāndogyopaniṣad* :

असद्वा इदमग्रासीत्। आत्मा वा इदमग्रासीद्। ब्रह्म वा इदमग्र आसीत्।

asadvā idamagrāsīt, ātmā vā idamagrāsīd, brahma vā idamagra āsīt

'That is, before the creation, *Prakṛti existed. Ātmā (Puruṣa)* existed and *Brahman* exited.'

2   Cf. following statements of *Upaniṣads* : *eko'haṁ bahusyām* 'I am alone, let me manifest in many.'

तदैक्षत् बहुस्याम् प्रजायेयेति

tadaikṣat bahusyām prajāyeyeti.

(He wished to be many, so gave birth to prajā'. So'kāmayat bahu syām prajāyeyeti: 'He wished to be many, so gave birth to people.'

See also Gitā: संकल्पप्रभवान् कामान् सर्वान् विद्धि (*Saṁkalpaprabhavān kāmān sarvān viddhi.* 'All things should be known to have been the creation of Will.'

3   सत्त्वरजस्तमसां साम्यावस्था प्रकृति

Sattvarajastamasāṁ sāmyāvasthā prakṛti (Sāṁkhya Darśana 1.1)

The process of evolution may be illustrated as under:

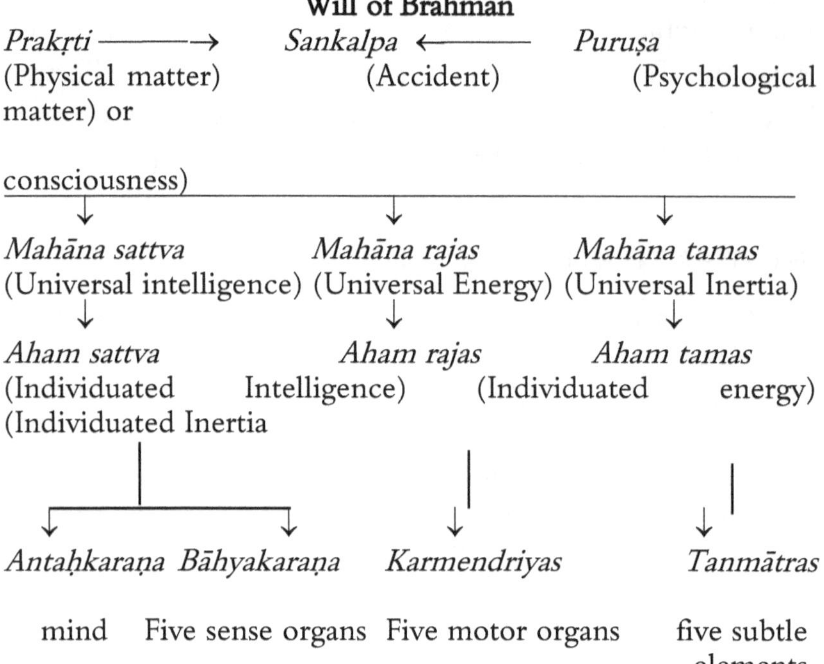

Thus based on the preceding discussion, it can be summed up that three types of elements play vital roles in the creation of the universe. For instance:

(1) The Prakṛti tattva, which evolves into tanmātras (pure bhūtas) and sthula bhūtas (gross or standard elements) are the upādāna (material) cause of the creation of living and non-living world, as the whole world is created as an outcome of the evolution of Prakṛti. Actually, Prakṛti tattva is the same, whether It is a living or non-living world. Perhaps due to the same reason it was theorised by ancient Ṛṣis (researchers): *yad aṇḍe tad brahmāṇḍe.* 'Whatever is contained in the

physiology is embodied in the physical world outside.[4] In other words, the Prakṛti within the living beings is closely attached to the Prakṛti without.

(2) Ātma tattva or Puruṣa Viśeṣa (Individuated soul) is the simple cause of the creation.

(3) Puruṣa tattva or Brahma (universal soul) is the efficient cause of the creation. Its accident with Prakṛti causes the evolution of the universe.

The above-mentioned Vedic philosophical concept of the world's creation points out three possible stages of evolution.

At the first stage, which may be called the stage of sensory perception, a philosopher conceives the world as merely a prākṛtika/material evolution due to ignorance (ajñānatā). This stage is known as the stage of utter ignorance. At this stage, his attempt at doing science will be the perfection of the physical and chemical properties of the Prakṛti / matter will evolve the material science known as positive sciences, viz. Physics and Chemistry. And based on the positive sciences, the type and nature of technology developed will be the one called Yāntrika, which pertains to Yantras/machines designed out of matter. In the second stage, apart from the utter ignorance stage, one can come out of ignorance and understand the role of Ātma tattva and Brahman in the creation of the world. One can also recognise Brahman's presence as an indifferent witness to the going about of the created world and recognises the proximity of Ātma tattva with that of the Prakṛti tattva within himself. At

---

4　*vide supra* 1.2.1

this stage, instead of doing with the outside Prakṛti, he tries to realize his true nature within.

This process is known as ātma-siddhi or self-perfection, or self-accomplishment, which can be achieved through Yoga, or say through regulation and concentration of mind (samādhi). The Yoga, which is also known as Ātma Yajña helps a Yogī in perfecting the antaḥ-prakṛti or nature within / bodily nature. Actually, there are two stages of samādhī, saṁprajñāta and asaṁprajñāta. At the saṁprajñāta level of samādhī, a Yogī maintains his separate identification or individuated existence and is not able to become one with the Almighty.

During this phase of his practice, a Yogī can, attain self-perfection or self-accomplishment, and this attainment, as it has already been said that the Prakṛti tattva is the same although regardless of the distinction of body or universe, accomplishes him to have a perfection over the Prakṛti without. At this stage, he develops extrasensory perception and is endowed with several divine powers on account of various perfection achieved through a regulated and concentrated mind e.g. divine vision, divine audition, divine olfaction, divine skin sensation, divine energy, divine motion, and speed, etc.[5] So at this stage, his sphere of science will be different from the material one. The concepts of light, heat, energy, sound, speed, motion, force, electricity, attraction, gravitation, etc., which are dealt with under material sciences or positive sciences, will change into the concepts to be dealt with under Bio-science.

---

[5] Cf. for details Patañjali's Yoga Darśanam

Now, he will deal with such concepts as bio-light, bio-heat, bio-energy, bio-sound, bio-speed, bio-mass, bio-force, bio-electricity, etc. Based on these concepts of Bio-science, he will (develop) a technology different from the one known to a nonprofessional. That will often appear to a layperson as a miraculous or magical one. For instance, television will find a replacement in the divine vision; telephone or telegraph Will be replaced by telepathy; different kind of transport, such as air, marine, and surface transports and road and rail technology will have their substitutes in the divine powers of utkrānti, jalāsaṅga, paṅkāsaṅga and kaṇṭakāsaṅga, etc.[6] as mentioned in the Yogadarśana.

Thus, the technology at this stage is known as the Tēntrika one.

The third and final stage is that of asaṁprajñāta samādhī in which a Yogī ultimately realises Brahman and a sense of duality is lost. It may be known as a stage of self-perception. It is a state of complete bliss, a state of complete union with God. It is a state described in Upaniṣads as:

अहं ब्रह्मास्मि

Ahaṁ brahmāsmi (I am Brahma)

तत्त्वमसि

Tattvamasi (Thou art He)

सर्वम् खल्विदम् ब्रह्म नेह नाना अस्ति किञ्चित्

Sarvam khalu idam Brahma neha nānā asti kiñcit

---

[6] उदान्-जयाज्जल-पङ्ककण्ड्कादिष्वासङ्ग उत्क्रान्तिश्च

udān-jayājjala-paṅkakaṇṭkādiṣvāsaṅga utkrāntiśca.

Yoga Darśana 3.39

(All that is visible is Brahma, nothing is here of a different kind).

At this non-duality stage, when a Yogī becomes virtually Brahman, he develops all such qualities as are thought to be the Godly ones. Thus the concepts of material science / positive science that evolved into the ideas of Bio-science in the second stage culminate in the third stage into the absolute ones, such as absolute light, absolute heat, absolute energy, and absolute mass, etc.

He, then, becomes able to do or materialise anything instantly at his will, like that of God himself. In other words, he deals with the science of the absolute ones. His technology finds an unending and infinite source in the form of absolutes. This source will remain unaffected by how much it is exploited. This fact finds a nice expression in the utterance of an Upaniṣadic Ṛṣi as:

पूर्णमदः पूर्णमिदम् पूर्णात् पुर्णमुदुच्यते
पूर्णस्य पूर्णमादाय पूर्णमेवावशिष्यते ॥

pūrṇamadaḥ pūrṇamidam pūrṇāt purṇamuducyate
pūrṇasya pūrṇamādāya pūrṇamevāvśiṣyate.

'That is, if absolute is taken out of absolute, the remaining will be absolute.'

The significance of this concept of absolutism has also been recognised in modern science as infinity (∞). But it finds its application ($\infty-\infty=\infty$) merely in theory rather than practice. I hope the further advancement of modern science will be able to discover the Bio-scientific values of the concepts that are dealt with in the positive sciences and will also be able to recognise the importance of infinity (∞) at a broader and larger scale of its applicability. The nature of this type of science and technology is described as Māntrika, which is more often than not enshrined in the four corners of adhyātma.

All these three types of science and technology found their recognition and codification in the whole Vedic literature. For instance, all the Vedic mantras signify adhyātma, adhidaivata, and adhibhūta, which further support the concept of Mantra, Tantra, and Yantra, respectively. The composition of the later Vedic literature, e.g. Brāhmaṇas, Āraṇyakas, Upaniṣads, and Sūtras was carried out keeping in view the above concept.

Six systems of Indian Philosophy also take up into consideration one or two subjects from the three concepts mentioned above of science, e.g. Vedānta or Uttar Mimāṁsā deals exclusively with the adhyātma; Vaiṣe "ika of Kaṇāda is an elaboration in positive sciences (adhibhūta); Sāṁkhya and Yoga cover an adhidaivata aspect; Nyāya takes up the methodology of doing science and Pūrva Mimaṁsā deals with the Yāntrika technology of ancient science. Similarly, three types of Vedic technology are mentioned in the books dealing with technology. For instance, Bṛhadvimāna "āstra, an ancient Vedic treatise on the science and technology of aeronautics, composed by Maharṣi Bhāradvāja makes a mention of three types of Vimānas (airplanes) as Māntrika, Tāntrika and Yāntrika ones devised on the basis of the science of Mantra (adhyātma), Tantra (adhidaivata) and Yantra (adhibhūta) respectively. Similarly, in the Dhanurveda, also we find the use of māntrika, tāntrika, and yāntrike weapons, like missiles and rockets.

The Development of Science and Technology in the Vedic Age

Although all the three types of science and technology were practised by the Vedic people, yet in the earliest phases of Vedic life, the science of mantra and the Māntrika technology was more practised by the high-spirited Vedic Ṛṣis. The term Ṛṣi itself signifies the researcher dealing with mantra's science. *ṛṣayo mantradṛṣṭāraḥ* or *ṛṣirdarśanāt.* 'Ṛṣis were those who had the perception of Mantra. Mantra or Śruti is the direct or primary knowledge acquired by the Ṛṣi or seer during self-perception. The mantra being related to the science of Brahman, was also known as Brahma. The concept of Śabda Brahma in the philosophy of grammar takes its shape from the above-quoted Vedic innovation of Mantra-Brahma. The Ṛṣis who knew or did the science of Mantra or Brahma were known as Brāhmaṇas. Brahma jānāti Brāhmaṇam. The technique of devising or developing the Māntrika technology based on the science of God/Mantra/Brahma was called the Brahma-Yajña. The above-mentioned science of Mantra/Brahma and the technique of Brahma Yajña was incorporated into the framework of adhyātma. As per traditional convention, all the Vedic Mantras primarily signify adhyātma, adhidaivata, and adhibhūta as the secondary significance. The Brāhmaṇas, which are the explanatory notes of the Vedas, refer to the sense of adhyātma 100 times, whereas adhidaivata and adhibhūta have been referred to 60 and 9 times, respectively. Adhyātma was the monopoly of a few high-spirited Ṛṣis, and its practice by the layman was not possible. Moreover, the seers never gave this technique to those who did not deserve this. So, in ancient times, the commoner remained aloof, unfamiliar, or unacquainted with this science's highly advanced and sophisticated nature for want of their interest in yogic science. It was a subjective concept that could not benefit those subjects who could not practise it. This is why, as is gathered from the stories of Epics and Purāṇas, the people who were keen to learn this science had to do penance to please such Ṛṣas were well equipped with this science to learn it from them. The word Smṛti is suggestive of the indirect or secondary knowledge that was made to realize by Brahmarṣis, having the Śruti knowledge, to later Ṛṣis.

On the contrary, in the later phases of Vedic life, owing to the decline in the high values and concepts of spirituality, the practice of oneness with God ceased to be observed and went out of vogue. This gave rise to the decline of Brahmarṣis tradition. The sorry fall of this noble tradition has also been alluded to by Yāska (2700 BC) as:

साक्षात्कृधर्माणः ऋषयो बभूवुः ॥ ते अवरेभयः असाक्षात्कृत्धर्मभ्यः उपदेशेन मन्त्रान् सम्प्रादुः

sākṣātkṛdharmāṇaḥ ṛṣayo babhūvuḥ. te avarebhayḥ asākṣātkṛtdharmabhyaḥ upadeśena mantrān samprāduḥ

"That is, there were born Ṛṣis to whom the science of Mantra (Dharma) was revealed. They preached this science by oral instructions to their juniors devoid of this revelation.'

With the degradation of the high values of spirituality and the decline of the noble tradition of Brahamrṣis, there arose a new generation of Ṛṣis whose concept of spirituality remained confined up to the observance of samādhī to its asamprajñāta level, i.e. the level at which they were endowed with divine powers or extra-sensory powers. Because they were endowed with only divinity, they were often called Devas. We come across numerous references in the Brāhmaṇas and Āraṇyakas where the Devas are differentiated from Manuṣyas on account of their being ūrdhvaretas. Thus the later Vedic epoch records the currency of the science of Tantra achieved through the technique of Ātma-Yajña. The science and technology based on Tantra being divine (daivika) in nature and appearance fall in ādhidaivika activities. But we find a further decline in spirituality, the people having been confined to the divine aspect (or siddhi) of Yoga. The attainment of siddhi or divine power of some sort or other is not very difficult to achieve through a bit of practice of Yoga. These siddhis or divine powers are described by Patañjali, the composer of Yoga Darśana as the hurdles on the way of real progress of the Yogī, as the Yogī often endowed with these powers and led astray from his actual path and is led by the mortal or worldly longings for name and fame. At this stage, a Yogī is described or christened as Bhraṣṭa Yogī or the fallen one.

This concept of Bhraṣṭa Yoga presents a distorted picture of the science and technology of Tantra in the post-Vedic period. The people are often seen to have indulged in the demonstration and performance of māyāvi śaktis or magical powers achieved sometime through Yoga and sometimes through the knowledge of specific chemical or physical properties of the material things which were unknown to both the learned or laity in the society. Numerous references to this regard of practices can be gathered from the Epics and Paurāṇika literature.

The increasing practice of the black rituals like that of Māraṇa (to kill someone), ucchāṭana (to trouble Someone), sammohan (to render someone unconscious), and vaśikaraṇa (to tame someone for his use) in the public life also exhibit a decline in the standard and distortion of the healthy tradition of science of Tantra. The present-day magicians and miracle men also represent this distorted form of Tāntrika tradition. Though the science of Tantra could find a wider sphere than the science of mantra, it could also not percolate to the commoner in society. So, the commoner remained aloof of the actual significance of the science of Tantra and thus deprived of the benefits which could be availed from the Tāntrika techniques. Moreover, he became the victim of clever tricks of black magic often played by the so-called Tāntrikas. This type of ritualism was responsible for the decline of Vedic values and the origin of Buddhism and Jainism in this country.

The material requirements also compelled the Vedic people to do the science of matter and develop and devise the Yāntrika technology known as Bhūtayajña.

Actually, in the post-Vedic period, when the science of mantra almost went out of vogue due to its subtleties and the science of Tantra also began to fizzle out owing to the distortion of its pure and original nature, it was only the Yāntrika science and technology that could eventually make its headway, perhaps, on account of its easily understandable nature. In the initial stages, though for the namesake, it assumed a separate and distinct form as a technology, in actual practice, it remained combined with the Tantra. It was practised under the influence of Tantra, so the Yāntrika technology was also operated with the help of some Mantra (thought in mind) and Tantra (sound, remote touch), etc. We come across such references to show the full development of this technology in the Epics, which mention the sophisticated type of weapons that were remotely controlled by some Mantras (thought in mind) or Tantra (sound or touch). This was known as the abhimantrita use of Yantra (machine/weapon). But with time Yāntrika science and technology assumed its pure form and no longer depended upon Tantra for its operation.

The extant Bṛhadvimānaśāstra and the Yantrasarvasva, quoted by the Bṛhadvimānaśāstra, are living testimony to the highly advanced nature of ancient Indian Yāntrika science and technology. Both of these treatises give a detailed break-up of scientific advancement and technological know-how in the field of aeronautics, machines, and other scientific instruments.

The Bṛhadvimānaśāstra[7] deals with the theory and operation of three types of Vimānas. Māntrika, Tāntrika, and Yāntrika could be operated on the land, in water, or the sky as needed. Based on Yantrasarvasva, the book discusses installing various instruments to save the aeroplane from multiple dangers. For instance, Śiraḥkilaka yantra is described as dispelling the effect of lightning. VarṣopsaŠhāra, Trāyasyavātanirsana, and ĀtapopasaŠhāra instruments to neutralise the effect of rain, winds, and heat respectively on the airplane. The

---

[7]   The manuscript of Bṛhadvimānaśāstra came to the notice of Maharshi Dayanand Saraswati (1824-1883) in 1875, as he makes a mention of this mauscript in one of his lectures delivered in Pune in 1875. All his lectures delivered in Pune in 1875 are compiled in the form of a book totled 'Poona Pravachana'.

extant Bṛhadvimānaśāstra is a book that validly answers the magical warfare often described to have been fought in ancient times. Moreover, the same book records the name of some forty personalities who were engaged in the science and technology of aeroplanes. A further breakthrough in the advancement of Yāntrika science and technology of ancient India is possible if and only if such extant literature on ancient sciences as Bṛhadvimānaśāstra and Yantrasarvasva is thoroughly and practically studied and other such literature as has been lost in the hoary past is searched out.

In the medieval period, yāntra was also applied in the esoteric sphere. In esoteric worship, the yantra was a chart that stored up within its confines spiritual power; drawn on a flat surface or made in relief, it had components or details which had to be strictly conformed to and, as an instrument for achieving spiritual power, it eminently deserved the name "yantra". In the minds of the writers, who were essentially men of spiritual yearning, the yantra or machine always suggested a highly apt analogy for the material universe or the mundane body activated by a God or presided over by a Soul.

3
Classification of Sciences and Technologies

In ancient India, sciences and arts (technologies) were also studied by 'Vidyā' and 'Kalā'. Although there was no end to sciences and technologies, broadly sciences were classified into 32 categories and arts or technologies into 64 categories in the *Sukranīti* (4.3.23) an ancient Indian text of 4th Century AD.

विद्या ह्यनन्ताश्च कलाः संख्यातुं नैव शक्यते ।
विद्यामुख्याश्च द्वात्रिंशच्चतुःषष्टिकलाः स्मृता ॥ 4.3.23

vidyā hyanantāścha kalāḥ saṁkhyātuṁ naiva śakyate /
vidyāmukhyāścha dvātriṁśachchatuḥṣaṣṭikalāḥ smṛtā //

According to the *Sukranīti* (4.3.24), vidyas are studied in theoretical form, and kalās are their applied aspect, so people who are deaf or hard of hearing and dumbs could also be the best technocrats, even though they were not able to teach effectively.

यद्यत्स्याद्वाचिकं सम्यक्कर्मविद्याभिसंज्ञकम् ।
शक्तो मूकोऽपियत्कर्तुं कलासंज्ञं तु तत्समृतम् ॥ शुक्रनीति, 4.3.24

yadyatsyādvāchikaṁ samyakkarmavidyābhisaṁjñakam /
śakto mūko'piyatkartuṁ kalāsaṁjñaṁ tu tatsamṛtam //

Hereunder we shall describe the nature and scope of 32 kinds of Vidyās and 64 kalās.

32 Types of Vidyās

Thirty-two types of Vidyās give a picture of

intellectual growth in India. 32 kinds of Vidyas classified into six categories are as under:

1. The First Category: In the first category, we have four Vedas– the *Ṛgveda*, the *Yajurveda*, the *Sāmaveda*, and the *Atharvaveda*. Hereunder we shall describe the nature and scope of 32 kinds of Vidyās and 64 kalās.

ऋग्यजुः साम चाथर्ववेदा आयुर्धनुः क्रमात्। शुक्रनीति, 4.3.25

ṛgyajuḥ sāma chātharvavedā āyurdhanuḥ kramāt /

Śukranīti, 4.3.25

2. The Second Category: In the second category, we have four Upavedas– Ayurveda (Upaveda of the *Ṛgveda*), Dhanurveda (Upaveda of the *Yajurveda*), Gāndharvaveda (Upaveda of the *Sāmaveda*), and Tantraveda or Arthaveda (Upaveda of Atharvaveda). Note: Arthaveda deals with rules of polity, economy, administration, and society).

गान्धर्वश्चैव तन्त्राणि उपवेदाः प्रकीर्तिताः ॥ शुक्रनीति, 4.3.25

gāndharvaschaiva tantrāṇi upavedāḥ prakīrtitāḥ ॥

Śukranīti, 4.3.25

3. The Third Category: In the third category, we have six Vedāṅgas– Śikṣā (Pronunciation), Vyākaraṇa (Grammar), Kalpa [Śrauta (Yajñas symbolizing the process of creation) and Smārta (16 Sanskāras) yajñas], Nirukta (Etymology), Jyotiṣa (Astronomy) and Chhandas (Metres).

शिक्षा व्याकरणं कल्पो निरूक्तं ज्यौतिषं तथा।
छन्दः षडंगानीमानि वेदानां कीर्तितानि हि ॥ शुक्रनीति, 4.3.26

śikṣā vyākaraṇaṁ kalpo nirūktaṁ jyautiṣaṁ tathā /
chhandaḥ ṣaḍaṁgānīmāni vedānāṁ kīrtitāni hi ॥

Śukranīti, 4.3.26

4. The Fourth Category:: In the fourth category we

have six Darśanas– Mīmānsā, Nyāya (Tarka), Sānkhya, Vedānta, and Yoga

5. **The Fifth Category:** In the fifth category, we have all texts dealing with the Itihāsa (History of creation), Purāṇa (History of creation, decreation, genealogies of Grahas and Nakṣatras, Time cycles), Smṛtis (Texts dealing with the rules and regulation of polity, economy, society, family, and individuals) and the ideology of atheists or non-believers.

मीमांसातर्कसांख्यानि वेदान्तो योग एव च ।
इतिहासाः पुराणानि स्मृतयो नास्तिकं मतम् ॥ शुक्रनीति, 4.3.28

mīmāṁsātarkasāṁkhyāni vedānto yoga ēva cha |
itihāsāḥ purāṇāni smṛtayo nāstikaṁ matam ||

Śukranīti, 4.3.28

6. **The Sixth Category:** In the sixth category are included Arthaśāstra, Kāmaśāstra, Ślpaśāstra (books dealing with technology and engineering), Alankāraśāstra, Kāvyas (literature), Vernaculars, Wise sayings suitable to various occasions, Yavana (people residing northwest frontier of India) philosophy and rule and regulations of various nations and countries (including his own).

अर्थशास्त्रं कामशास्त्रं तथा शिल्पमलंकृतिः ।
काव्यानि देशभाषाऽवसरोक्तिर्यावनं मतम् ॥ शुक्रनीति, 4.3.29
देशादिधर्मा द्वात्रिंशदेता विद्याभिसंज्ञिताः ॥ शुक्रनीति, 4.3.30

arthaśāstraṁ kāmaśāstraṁ tathā śilpamalaṁkṛtiḥ |
kāvyāni deśabhāṣā'vasaroktiryāvanaṁ matam ||
deśādidharmā dvātriṁśadetā vidyābhisaṁjñitāḥ ||

Śukranīti, 4.3.29-30

64 Types of Kalās

64 Types of kalās give a picture of the industrial and

economic growth of the people in ancient India. The picture of Vidyās and Kalās together constitutes a graphic account of the social life of the people of India who lived in ancient times.

After going through this, one can hardly believe that ancient Indians were a race of abstract metaphysicians who were negligent of the actual needs of society and cultivated the art of preparing for the next life only. One would be given to understand that ancient Indians knew how to enjoy life and supply its necessities, comforts, and decencies in an eco-friendly manner. Economically speaking, they were as self-sufficient as any people could be and made their material and secular life as comfortable and happy as possible. Intellectually speaking, they were competent enough to investigate not only the highest truths of the universe-the eternal problems of existence but also to study and discuss all those branches of learning which had for their aim the practical furtherance of social ends-the amelioration of human life.

The more one studies the social, economic, political, and other secular facts of the civilization of the ancient Indians, the more one is impressed with the fact that their institutions -industrial, educational, and administrative-were adequate for all the ends of human existence; and if they differ from anything of the kind in modern times or other countries it is because of the adaptation to the circumstances and conditions of time and place which is the fundamental cause of all varieties and divergences in the universe. Those who advocate the doctrine of the relativity of institutions cannot think of the Hindus as an economically inefficient or politically incompetent race or as one with no industrial or political aptitudes. The fact instead is, in all these aspects of secular life, they represent a distinct type that is not necessarily low, medieval, or primitive simply because it does not resemble the predominant types today.

Now hereunder, we give a count of 64 types of kalās. It may be pointed out that out of 64 kinds of kalās, 23 are derived from the 4 Upavedas. For instance,

From Gāndharvaveda are derived the kalās of

1. Dancing with appropriate gestures and movements.

2. Playing musical instruments.

3. Decoration of men and women by dress and ornaments.

4. Knowledge of sundry mimicry and antics.

5. Laying our beds and furniture and weaving garlands.

6. Entertainment of people by gambling and various tricks of magic.

7. Knowledge of different aspects of giving pleasure.

हावभावादिसंयुक्तं नर्तनं तु कला स्मृता ।
अनेकवाद्यविकृतौ ज्ञानं तद्वादनं कला ॥ 4.3.67
वस्त्रालंकारसंधानं स्त्रीपुंसोश्च कला स्मृता ।
अनेकरूपाविर्भावविकृतिज्ञानं कला स्मृता ॥ 4.3.68
शय्यास्तरणं संयोगपुष्पादिग्रथनं कला ।
द्यूताद्यनेकक्रीडाभीरंजनं तु कला स्मृता ॥ 4.3.69
अनेकासनसंधानै रतेज्ञानि कला स्मृता ।
कलासप्तकमेतद्धि गान्धर्वे समुदाहृतम् ॥ 4.3.70

hāvabhāvādisaṁyuktaṁ nartanaṁ tu kalā smṛtā |
anekavādyavikṛtau jñānaṁ tadvādanaṁ kalā || 67

vastrālamkārasamdhānam strīpumsoścha kalā smṛtā |
anekarūpāvirbhāvakṛtijñānam kalā smṛtā || 68

śayyāstaraṇam samyogapuṣpādigrathanam kalā |
dyūtādyanekakrīḍābhīramjanam tu kalā smṛtā || 69

anekāsanasamdhānai raterjñāne kalā smṛtā |
kalāsaptakametaddhi gāndharve samudāhṛtam || 70

From Āyurveda are derived the following 10 kalās:

8. Honey gathering, distillation and filtration technology.

9. Extrication of thorns and operations.

10. Food technology and Cooking of various dishes by intermix of various tastes.

11. Agriculture, horticulture, and sericulture.

12. The technology of stone melting and powdering and mixing with metals.

13. Sugar making technology.

14. Pharmacology and pharmacy.

15. Metallurgy.

16. Alloy making.

17. Extraction and preparation of various salts and alkalis, chemical process and manipulations.

मकरन्दासवादीनां मद्यादीनां कृतिः कला ।
शल्यगूढाहृतैः ज्ञानं शिरात्रणव्याधे कला ॥ 4.3.71

हीनादिरससंयोगान्नादिसंपाचनं कला ।
वृक्षादिप्रसवारोपपालनादिकृतिः कला ॥ 4.3.72

पाषाणधात्वादिद्रतिस्तऽस्मरणं कला ।
यावदिक्षुविकाराणं कृतिज्ञानं कला स्मृता ॥ 4.3.73

धात्वौषधीनां संयोग-क्रियाज्ञानं कला स्मृता ।
धातुसांकर्यपार्थक्य-करणं तु कला स्मृता ॥ 4.3.74

संयोगपूर्वविज्ञानं धात्वादीनां कला स्मृता ।
क्षारनिष्कासनज्ञानं कलासंज्ञं तु तत्स्मृतम् ॥ 4.3.75

कलादशकमेतद्धि ह्यायुर्वेदागमेषु च । 4.3.76

makarandāsavādīnāṁ madyādīnāṁ kṛtiḥ kalā /
śalyagūḍhāhṛtaiḥ jñānaṁ śirāvraṇavyādhe kalā // 4.3.71

hīnādirasasaṁyogānnādisaṁpāchanaṁ kalā /
vṛkṣādiprasavāropapālanādikṛtiḥ kalā // 4.3.72

pāṣāṇadhātvādidratista'smaraṇaṁ kalā /
yāvadikṣuvikārāṇaṁ kṛtijñānaṁ kalā smṛtā // 4.3.73

dhātvauṣadhīnāṁ saṁyoga-kriyājñānaṁ kalā smṛtā /
dhātusāṁkaryapārthakya-karaṇaṁ tu kalā smṛtā // 4.3.74

saṁyogapūrvavijñānaṁ dhātvādīnāṁ kalā smṛtā /
kṣāraniṣkāsanajñānaṁ kalāsaṁjñaṁ tu
*tatsmṛtam //*4.3.75

kalādaśakametaddhi hyāyurvedāgameṣu cha / 4.3.76

From Dhanurveda, following kalās developed–

18. Manufacturing of arms and their employment.

19. Duel fight, hand to hand fight, and marking of aim. An attack by duellers, that which is made by various dangerous artifices of hands, and by throwing down the opponent in various ways, etc. as well as technique to extricate oneself from these, like techniques of boxing.

20. The shooting of arms, missiles, and rockets towards some fixed point.

21. Formation of battle arrays according to the signals

given by musical instruments (bugles).

22. Employment of horses, elephants, chariots, tanks, cannons etc.

शस्त्रसन्धानविक्षेपः पदादिन्यासतः कला ॥ 4.3.76

संध्याघाताकृष्टिर्भेदैर्मल्लयुद्धं कला स्मृता ।
बाहुयुद्धन्तु मल्लानामशस्त्रं मुष्टिभिः स्मृतम् ॥ 4.3.77

मृतस्य तस्य न स्वर्गो यषो नेहापि विद्यते ।
बलदर्पविनाशान्तं नियुद्धं यशसे रिपोः ॥ 4.3.78

न कस्यासीत् कुर्याद् वा प्राणान्तं बाहुयुद्धकम् ।
कृतप्रतिकृतैश्चित्रैर्बाहुभिश्च सुसंकटैः ॥ 4.3.79

सन्निपातावघातैश्च प्रमोदन्मथनैस्तथा ।
कृतं निपीडनं ज्ञेयं तन्मुक्तिस्तु प्रतिक्रिया ॥ 4.3.80

कलाऽभिलक्षिते देशे यन्त्राद्यस्त्रनिपातनम् ।
वाद्यसंकेततो व्यूहरचनादि कला स्मृता ॥ 4.3.81

गजाश्वरथगत्या तु युद्धसंयोजनं कला ।
कलापंचकमेतद्धि धनुर्वेदागमे स्थितम् ॥ 4.3.82

śastrasandhānavikṣepaḥ padādinyāsataḥ kalā // 4.3.76

*saṁdhyāghātākṛṣṭirbhedairmallayuddhaṁ kalā smṛtā |
bāhuyuddhantu mallānāmaśastraṁ muṣṭibhiḥ
smṛtam* // 4.3.77

*mṛtasya tasya na svargo yaśo nehāpi vidyate |
baladarpavināśāntaṁ niyuddhaṁ yaśase ripoḥ* // 4.3.78

*na kasyāsīt kuryād vā prāṇāntaṁ bāhuyuddhakam |
kṛtapratikṛtaiśchitrairbāhubhiścha susaṁkaṭaiḥ* // 4.3.79

*sannipātāvaghātaiścha pramodanmathanaistathā |
kṛtaṁ nipīḍanaṁ jñeyaṁ tanmuktistu pratikriyā //
4.3.80*

*kalā'bhilakṣite deśe yantrādyastranipātanam |
vādyasaṁketato vyūharachanādi kalā smṛtā* // 4.3.81

gajāśvarathagatyā tu yuddhasaṁyojanaṁ kalā ǀ
kalāpaṁchakametaddhi dhanurvedāgame sthitam ǁ *82*

In addition to the above, following 42 types kalās were also developed in ancient India.

23. Sitting in various postures for doing yoga and yogic exercises.

विविधासनमुद्राभिर्देवतातोषणं कला। 4.3.83

vividhāsanamudrābhirdevatātoṣaṇaṁ kalā ǀ 4.3.83

24. The act of driving horses and elephants and training them.

सारथ्यं च गजाश्वादेर्गतिशिक्षा कला स्मृता ॥ 4.3.83

sārathyaṁ cha gajāśvādergatiśikṣā kalā smṛtā ǁ *4.3.83*

25. Manufacturing of earthenwares (ceramics), wooden, stone, and metal vessels. and four other kalās associated with them like

26. Cleansing of earthen, wooden, stone, and metal vessels.

27. Polishing earthen, wooden, stone and metal vessels

28. Dyeing or rinsing earthen, wooden, stone, and metal vessels.

29. Picture drawing and carving of earthen, wooden, stone and metal vessels

मृत्तिकाकाष्ठपाषाणधातुभाण्डादिसत्क्रिया कला।
पृथक्कलाचतुष्कन्तु चित्राद्यलेखनं कला ॥ 4.3.84

mṛttikākāṣṭhapāṣāṇadhātubhāṇḍādisatkriyā kalā ǀ
pṛthakkalāchatuṣkantu chitrādyalekhanaṁ kalā ǁ

4.3.84

30. Civil engineering, architecture, and sculpture (construction of houses, tanks, canals; leveling and road building)

तडागवापीप्रासादसमभूमिक्रिया कला । 4.3.85

taḍāgavāpīprāsādasamabhūmikriyā kalā | 4.3.85

31. Instrumentation or manufacturing of clocks, watches, musical, and several other instruments.

घट्याद्यनेकयन्त्राणां वाद्यानां तु कृतिः कला ॥ 4.3.85

ghaṭyādyanekayantrāṇāṁ vādyānāṁ tu kṛtiḥ kalā ॥

4.3.85

32. Dyeing of clothes with light, moderate and sharp colours).

हीनमध्यादिसंयोगनिरोधैश्च क्रिया कला । 4.3.86

hīnamadhyādisaṁyoganirodhaiścha kriyā kalā | 4.3.86

33. Manufacturing several machines by controlling and manipulating water, air, heat, and electricity.

34. Putting down fire, checking storms and floods

जलवाय्वग्निसंयोगनिरोधैश्च क्रिया कला ॥ 4.3.86

jalavāyvagnisaṁyoganirodhaiścha kriyā kalā ॥ 4.3.86

35. Marine engineering, road transport and aeronautics (manufacturing seafaring vessels, vehicle plying on roads and aeroplanes, etc.)

नौकारथादियानानां कृतिज्ञानं कला स्मृता । 4.3.87

naukārathādiyānānāṁ kṛtijñānaṁ kalā smṛtā | 4.3.87

36. Textile engineering: thread and rope making.

सूत्ररज्जुकरणं विज्ञानं तु कला स्मृता ॥ 4.3.87

sutrarajjukaraṇaṁ vijñānaṁ tu kalā smṛtā ॥ 4.3.87

37. Textile engineering: the weaving of fabrics and embroidering.

अनेकतन्तुसंयोगैः पटबन्धः कला स्मृता । *4.3.88*

anekatantusaṁyogaiḥ paṭabandhaḥ kalā smṛtā | 4.3.88

38. Marking of holes in ears and nose.

39. Testing of gems with regard to their quality.

वेधादिसदसंज्ञानं रत्नानां च कला स्मृता ॥ *4.3.88*

vedhādisadasaṁjñānaṁ ratnānāṁ cha kalā smṛtā ||

4.3.88

40. Testing of Gold and other metals for their purity.

स्वर्णादीनां तु याथात्म्य-विज्ञानं च कला स्मृता । *4.3.89*

svarṇādīnāṁ tu yāthātmya-vijñānaṁ cha kalā smṛtā |

4.3.89

41. Preparation of artificial Gold and gems.

कृत्रिमस्वर्णरत्नादि-क्रियाज्ञानं कला स्मृता ॥ *4.3.89*

kṛtrimasvarṇaratnādi-kriyājñānaṁ kalā smṛtā || 4.3.89

42. Making of ornaments with gold, silver, and other metals.

43. The enameling of various metals

स्वर्णाद्यलंकारकृतिः कला लेपादिसत्कृतिः । *4.3.90*

svarṇādyalaṁkārakṛtiḥ kalā lepādisatkṛtiḥ | 4.3.90

44. The softening of leathers.

मार्दवादिक्रियाज्ञानं चर्मणा तु कला स्मृता ॥ *4.3.90*

mārdavādikriyājñānaṁ charmaṇā tu kalā smṛtā ||

4.3.90

45. Flying of skins from the bodies of the dead animals.

पशुचर्मांगनिर्हार-क्रियाज्ञानं कला स्मृता । 4.3.91

paśucharmāṁganirhāra-kriyājñānaṁ kalā smṛtā ।

4.3.91

46. Milking.

47. Churning and clarification of butter.

दुग्धदोहादिविज्ञानं घृतान्तन्तु कला स्मृता ॥ 4.3.91

dugdhadohādivijñānaṁ ghṛtāntantu kalā smṛtā ॥

4.3.91

48. Tailoring.

सीवने कंचुकादिनां विज्ञानं तु कला स्मृता । 4.3.92

sīvane kaṁchukādināṁ vijñānaṁ tu kalā smṛtā ।

4.3.92

49. Swimming.

बाह्वादिभिश्च तरणं कलासंज्ञं जले स्मृतम् ॥ 4.3.92

bāhvādibhiścha taraṇaṁ kalāsaṁjñaṁ jale smṛtam ॥

4.3.92

50. Cleansing of domestic utensils.

मार्जने गृहभाण्डादेर्विज्ञानं तु कला स्मृता । 4.3.93

mārjane gṛhabhāṇḍādervijñānaṁ tu kalā smṛtā ।

4.3.93

51. Drycleaning and ironing of clothes.

52. Haircutting and shaving.

वस्त्रसंमार्जनं चैव क्षुरकर्म कले ह्युभे ॥ 4.3.93

vastrasaṁmārjanaṁ chaiva kṣurakarma kale hyu bhe ॥ 4.3.93

53. Extraction of and preparation of oils from seeds and fats.

तिलमांसादिस्नेहानां कला निष्कासने कृतिः । 4.3.94

tilamāṁsādisnehānāṁ kalā niṣkāsane kṛtiḥḥ । 4.3.94

54. Ploughing and climbing on trees.

सीताद्यकर्षणे ज्ञानं वृक्षाद्यारोहणे कला ॥ 4.3.94

sītādyakarṣaṇe jñānaṁ vṛkṣādyārohaṇe kalā ॥ 4.3.94

55. Knowledge of work in such a way as to please somebody.

मनोऽनुकूलयासेवायाः कृतिज्ञानं कला स्मृता । 4.3.95

mano'nukūlayāsevāyāḥ kṛtijñānaṁ kalā smṛtā । 4.3.95

56. Manufacturing of vessels with bamboo straws

वेणुतृणादिपात्राणां कृतिज्ञानं कला स्मृता ॥ 4.3.95

veṇutṛṇādipātrāṇāṁ kṛtijñānaṁ kalā smṛtā ॥ 4.3.95

57. Manufacturing of glass vessels and other items.

काचपात्रादिकरणविज्ञानं तु कला स्मृता । 4.3.96

kāchapātrādikaraṇavijñānaṁ tu kalā smṛtā । 4.3.96

58. Pumping and withdrawing of water.

संसेचनं संहरणं जलानां तु कला स्मृता ॥ 4.3.96

saṁsechanaṁ saṁharaṇaṁ jalānāṁ tu kalā smṛtā ॥

4.3.96

59. Manufacturing of various Śastras (tools, instruments, and machines) and Astras (missiles, rocket launcher, etc.)

लोहाभिसारशस्त्रास्त्रकृतिज्ञानं कला स्मृताः । 4.3.97

lohābhisāraśastrāstrakṛtijñānaṁ kalā smṛtāḥ । 4.3.97

60. Manufacturing of saddles for horses, elephants, bulls, and camels.

गजाश्ववृषभोष्ट्राणां पल्यादिक्रिया कला ॥ 4.3.97

gajāśvavṛṣabhoṣṭrāṇāṁ palyādikriyā kalā ǁ *4.3.97*

61. Nursing, caring, and education of children

शिशोः संरक्षणे ज्ञानं धारणे क्रीडने कला । 4.3.98

śiśoḥ saṁrakṣaṇe jñānaṁ dhāraṇe krīḍane kalā ǀ

4.3.98

62. Proper punishment to offenders for their various offences.

सुयुक्तताडनज्ञानमपराधिजने कला ǁ 4.3.98

suyuktatāḍanajñānamaparādhijane kalā ǁ *4.3.98*

63. Writing skills of various scripts of languages of various countries.

नानादेशीयवर्णानां सुसम्यग्लेखनं कला ǁ 4.3.99

nānādeśīyavarṇānāṁ susamyaglekhanaṁ kalā ǁ *4.3.99*

64. Making and preservation of betels.

ताम्बूलरक्षादिविज्ञानं तु कला स्मृता ǁ 4.3.99

tāmbūlarakṣādivijñānaṁ tu kalā smṛtā ǁ *4.3.99*

4

Civil Engineering

Building Construction

Vedic seers developed an advanced system in civil engineering. There are references to the planning and construction of buildings - residential, royal, military, and religious. Several references to architecture are found in Vedas, Sthāpatya Veda, and post-Vedic Sanskrit literature like Vāstu Śāstra, Maya-madam, Mānasāra, Samarāṅgaṇa Sūtradhāra, Purāṇas, Bṛhat Saṁhitā, and Arthasāstra. The Śālā Nirmāṇa Sūkta' (3.12) and 'Śālā Sūkta' (9.3)[8] of Atharvaveda states to construct a beautiful, well-designed, spacious house with doors on all sides.

As per Atharvaveda (AV.) 9.3.7), a good house must have provisions of light and provisions of rooms for the kitchen (agniśālā), stores, Yajña śālā (havirdhāna for performing havana), entertaining guests (sada), ladies to sit and rest (patnīnāṁ sada), learned persons to talk and

---

8 उपमिता प्रतिमितामथो परिमितामुप ।
शालाया विश्ववाराया नद्धानि वि चृतामसि ॥

upamitā pratimitām atho parimitām up

śālāya viśvavārāyā naddhāni vicṛtām asi.

'Let us construct a beautiful, well designed, commodious house. Let us strengthen the ties and fastenings of the house that has doors on all sides and holds all facilities.'

discuss various topics (devanaṁ sada). It should have a lighting system.

हविर्धानमग्निशालं पत्नीनां सदनं सदः ।
सदो देवानामसि देवि शाले ॥ 9.3.7 ॥

*havirdhānamagniśālaṁ patnīnāṁ sadanaṁ sadaḥ /
sado devānāmasi devi śāle // 9.3.7*

It should be well designed and well ventilated (AV. 9.3.8) and well decorated (AV. 9.3.10)

अक्षुमोपशं विततं सहस्राक्षं विषूवति ।
अवनद्धमभिहितं ब्रह्मणा वि चृतामसि ॥9.3.8

अमुत्रैनमा गच्छताद् दृढा नद्धा परिष्कृता ।
यस्यास्ते विचृतामस्यङ्गमङ्गं परुष्परुः ॥9.3.10

*akṣumopaśaṁ vitataṁ sahasrākṣaṁ viṣūvati /
avanaddhamabhihitaṁ brahmaṇā vi chṛtāmasi // 9.3.8*

*amutrainamā gachchhatād dṛḍhā naddhā pariṣkṛtā /
yasyāste vichṛtāmasyaṅgamaṅgaṁ paruṣparuḥ // 9.3.10*

During the Vedic period, houses were made of stones and wood (AV. 9.3.12).

नमस्तस्मै नमो दात्रे शालापतये च कृण्मः ।
नमोऽग्नये प्रचरते पुरुषाय च ते नमः ॥ *9.3.12*

*namastasmai namo dātre śālāpataye cha kṛṇmaḥ /
namo'gnaye pracharate puruṣāya cha te namaḥ // 9.3.12*

Houses also sheltered domestic animals like cow, horse, etc. (AV. 9.3.13-14)

गोभ्यो अश्वेभ्यो नमो यच्छालायां विजायते ।
विजावति प्रजावति वि ते पाशांश्चृतामसि ॥9.3.13 ॥

अग्निमन्तश्छादयसि पुरुषान् पशुभिः सह ।
विजावति प्रजावति वि ते पाशांश्चृतामसि ॥9.3.14 ॥

gobhyo aśvebhyo namo yachchhālāyāṁ vijāyate /

vijāvati prajāvati vi te pāśaṁśchṛtāmasi // *9.3.13*

agnimantaśchhādayasi puruṣān paśubhiḥ saha /
vijāvati prajāvati vi te pāśaṁśchṛtāmasi // *9.3.14*

Houses were founded and build upon a nice plot. They were rich in prosperity, milk, and water. (AV. 9.3.16)

ऊर्जस्वती पयस्वती पृथिव्यां निमिता मिता ।
विश्वान्नं विभ्रती शाले मा हिंसीः प्रतिगृह्णतः ॥9.3.16

ūrjasvatī payasvatī pṛthivyāṁ nimitā mitā /
viśvānnaṁ vibhratī śāle mā hiṁsīḥ pratigṛhṇataḥ //9.3.16

The doors were bolted. (AV. 9.3.18)

इटस्य ते बि चृताम्यपि नद्धमपोर्णुवन् ।
वरुणेन समुब्जितां मित्रः प्रातर्व्युऽब्जतु ॥9.3.18

iṭasya te bi chṛtāmyapi naddhamapornuvan /
varuṇena samubjitāṁ mitraḥ prātarvyu'bjatu // *9.3.18*

Homes were made proof of air, fire, and other natural forces. Houses were planned intelligently, built and erected by skilled, expert engineers. (AV. 9.3.19)

ब्रह्मणा शालां निमितां कविभिर्निमितां मिताम् ।
इन्द्राग्नी रक्षतां शालाममृतौ सौम्यं सदः ॥ 9.3.19

brahmaṇā śālāṁ nimitāṁ kavibhirnimitāṁ mitām /
indrāgnī rakṣatāṁ śālāmamṛtau saumyaṁ sadaḥ //9.3.19

The houses had two, four, six, eight, or ten rooms. There was no provision for odd nos. of rooms in the house (AV. 9.3.21).

या द्विपक्षा चतुष्पक्षा षड्क्षा या निमीयते ।
अष्टापक्षां दशपक्षां शालां मानस्य पत्नीमग्निर्गर्भइवा शये ॥ 9.3.21

yā dvipakṣā chatuṣpakṣā ṣaṭpakṣā yā nimīyate /
aṣṭāpakṣāṁ daśapakṣāṁ śālāṁ mānasya

patnīmagnirgarbhaivā śaye // 9.3.21

There was the provision of multistoried buildings/flats where rooms were built upon rooms, and families used to live on different floors (AV. 9.3.20).

कुलायेऽधि कुलायं कोशे कोशः समुब्जितः ।
तत्र मर्तो वि जायते यस्माद् विश्वं प्रजायते ॥ 9.3.20

kulāye'dhi kulāyam kośe kośaḥ samubjitaḥ /
tatra marto vi jāyate yasmād viśvam prajāyate // 9.3.20

Houses were built in the west, i.e. east facing so that the house owner may enter the house with his face towards the west. The provision of fire and water in the place was ensured, as they were considered the main doors of life (AV. 9.3.22).

प्रतीचीं त्वा प्रतीचीनः शाले प्रैम्यहिंसतीम् ।
अग्निर्ह्यऽन्तरापश्चर्तस्य प्रथमा द्वाः ॥ 9.3.22 ॥

pratīchīm tvā pratīchīnaḥ śāle praimyahimsatīm /
agnirhya'ntarāpaśchartasya prathamā dvāḥ // 9.3.22

Purified, potable drinking water was supplied in the houses (AV. 9.3.23).

इमा आपः प्र भराम्ययक्ष्मा यक्ष्मनाशनीः ।
गृहानुप प्र सीदाम्यमृतेन सहाग्निना ॥ 9.3.23 ॥

imā āpaḥ pra bharāmyayakṣamā yakṣamanāśanīḥ /
gṛhānupa pra sīdāmyamṛtena sahāgninā // 9.3.23

There was also a provision of lightweight, portable houses that could be carried/shifted at will to accompany the house owner like that of the house lady (AV. 9.3.24).

मा नः पाशं प्रति मुचो गुरुर्भारो लघुर्भव ।
वधूमिव त्वा शाले यत्र कामं भरामसि ॥ 9.3.24

mā naḥ pāśam prati mucho gururbhāro laghurbhava /

vadhūmiva tvā śāle yatra kāmaṁ bharāmasi // 9.3.24

Soil Testing

Soil testing was a common practice in ancient India before construction. The *Matsya Purāṇa* (253.11)[9] makes a provision for soil testing.

Land with anthills, skeletons, full of pits, and craters was be often avoided. We find its mention in Vāstu Śāstra (śloka, 5)[10].

The building material was used according to the colour, smell, taste, shape, sound, and touch of the soil detected on examination (*Bhṛgu Saṁhitā,* chapter, 4)[11].

For laying the foundation, the earth was to be dug till the water level.[12]

---

9   पूर्वं भूमिं परिक्षेत पश्चात् वास्तु प्रकल्पयेत् ।
pūrvaṁ bhūmiṁ parikṣet paścāt vāstu prakalpayet
'First test the site, after that plan the construction.'

10   वल्मीकेन समायुक्ता भूमिरस्थिगणैस्तु या ।
रन्ध्रान्विता च भूर्वर्ज्या गतीघेष्च समन्विता ॥
valmikena samāyuktā bhūmirasthi gaṇaistu yā
randhrāuvitā ca bhūrvajyāgatī gheśca samanvitā.
'Land with anthils, skeletons, full of pits and craters should be avoided.'

11   वर्णगन्धरसाकारादिशब्दस्पर्शनैरपि ।
परिक्ष्यैव यथायोग्यं गृह्णियाद् द्रव्यमुत्तमम् ॥
varṇa-gandha-rasa-ākāra-ādiśabda sparśanairapi
parikṣyaiva yathāyogyaṁ gṛhṇiyād dravyamuttamam
'After examining the colour, smell, taste, shape, sound and touch of the soil buy best material as found suitable.'

12   यावत्तत्र जलं दृष्टं खनेत्तावत्तु भूतले ॥
yāvattatra jalaṁ dṛṣṭaṁ khanettāvattu bhūtale
'Till water is seen there, one should dig the ground.'

The soil was tested by digging a pit of one arm's length and refilling it with the same ground; if the earth is more, one will spawn prosperity; if short, one will generate loss, if equal, it is normal (*Kāśyapa Saṁhitā*, chapter 4)[13].

It was a very scientific way of testing. If some soil is left after refilling the pit, it shows the scope for compaction of the sand, and such soil is considered desirable. This testing procedure mentioned in ancient Sanskrit texts is in vogue to this day.

Making of bricks

Vedic people were familiar with the technology of brickmaking. They made bricks that could last not years but a thousand years. The altars were built of bricks.

For making bricks, proper clay was chosen. The clay was classified into four kinds: salty, off-white, smooth black, and coarse red. Clay devoid of gravel, pebbles, roots, and bones and soft in touch was considered suitable for brick making. (Maya-Matam, Kalā-Mūla-śāstram by Kapila Vātsyāyana (śloka 114-115)[14].

---

13   रत्निमात्रमधे गर्तं परीक्ष्य खातपूरणे ।
अधिके श्रियामाप्नोति न्यूने हानिं समे समम् ॥
ratnimātramadhe garte parīkṣya khātapūraṇe
adhike śriyamāpnoti nyune hāniṁ same samam.
'Soil should be tested by digging a pit of one arm length and refilling it with the same soil. If soil is more, one will beget prosperity, if short, one will beget loss; if equal, it is normal.'

14   ऊषरं पाण्डुरं कृष्णचिक्कणं ताम्रपुल्लकम् ।
मृदश्चतस्रस्तास्वेव गृह्णीयात् ताम्रपुल्लकम् । 114
ūṣaraṁ pāṇḍuraṁ kṛṣṇachikkaṇaṁ tāmrapullakam /
mṛdaśchatasrastāsveva gṛhṇīyāt tāmrapullakam / 114

For making bricks, the clods of suitable clay were filled in knee-deep water; then, it was mixed and pounded with the feet forty times repeatedly. (Maya-Matam, Kalā-Mūla-śāstram by Kapila Vātsyāyana, (śloka 116)[15].

The clay was also soaked in the sap of fig, kadamba, mango, abhaya, and akṣa and in Triphala's water for three months before pounding. (Maya-Matam, Kalā-Mūla-śāstram by Kapila Vātsyāyana (śloka, 117)[16].

These bricks were made in four, five, six, eight, and unit widths and twice in length. Their depth in the middle and the two ends is one-fourth or one-third the width. Again these bricks were typically dried and baked for one, two, three, or four months. Their usability test was again done by soaking them in the water; they were declared suitable for use after passing the test. (Maya-Matam, Kalā-Mūla-śāstram by Kapil done by soaking them in the water; they were declared suitable for use after passing the test. (Maya-Matam, Kalā-Mūla-śāstram by Kapila Vātsyāyana (śloka Vātsyāyana (śloka, 118-120)[17].

अशर्कराश्ममूलास्थिलोष्टं सतनुवालुकम् ।
एकवर्णं सुखस्पर्शमिष्टंलोष्टेष्टकादिषु ॥ 115

aśarkarāśmamūlāsthiloṣṭaṁ satanuvālukam /
ēkavarṇaṁ sukhasparśamiṣṭaṁloṣṭeṣṭakādiṣu // 115

15   मृत्खण्डंपूरयेदग्रे जानुदघ्ने जले ततः ।
आलोड्य मर्दयेत् पद्‍भ्यां चत्वारिंशत् पुनः पुनः । 116

16   क्षीरद्रुम कदम्बाम्राभयाक्षत्वञ्जलैरपि ।
त्रिफलाम्बुभिरासित्त्वा मर्दयेन्मासमात्रकम् ॥ 117

17   चतुष्म०चषडष्टाभिर्मात्रैस्तद्द्विद्विगुणायता । 118
व्यासार्धार्धत्रिभगैकतीव्रा मध्ये परेऽपरे ।
इष्टकाबहुषः शोष्याः समदग्धाः पुनष्च ताः ॥ 119
एकद्वित्रिचतुर्माससमतीत्यैव विचाक्षणः ।

Types of Cements, Mortars, and Lutes

India is full of buildings constructed hundreds and thousands of years ago depicting the skill of civil engineering throughout the length and breadth of the country. These structures of civil engineering still exist withstanding the ravages of time. It is natural to question how stones and bricks were made to bind during that period. How were the hefty wooden doors, windows, and frames secured? What kind of binders were prepared and used to cement them together? The answers to these questions can be obtained from the ancient Vedic and Post-Vedic Sanskrit literature. Bṛhat Saṁhitā (Chapter 56) has one whole chapter dedicated to this particular topic titled "Vajralepa lakṣṇādhyāya" meaning "Adamantine Glue". These glues called Aṣṭabandha are still prepared in temple premises for fixing or refixing Murtis. Some of the formulae are cited as under:

Take the unripe fruits of tinduka (Diospyros paniculata) and kapithala (Feronia elephantum), flowers of silk cotton (Morus acedosa), seeds of shallaki (Boswellia serrata), bark of Dhanvana, and vacā (Orris root); boil all of them in a droṇa (5 sheer or 256 palas) water and reduce the decoction to the 8th part of its original volume (i.e. 32 palas). Mix the paste with the following substances, viz. śrivasaka (a secretion of a tree used as incense, turpentine?), rasa (myrrh), guggulu (Commiphora roxburghii), Bhallataka (Semecarpus anacardium), Kunduruka (cunduru, secretion of Deodar), sarjarasa (secretion of resin), atasī (Linum

---

जले प्रक्षिप्य यत्नेन जलादुद्धृत्य तत् पुनः ॥ 120

ēkadvitrichaturmāsasamatītyaiva vichākṣaṇaḥ |
jale prakṣipya yatnena jalāduddhṛtya tat punaḥ ǁ 120

usikatissimum) and Bilva fruit (Aegle marmelos). The resulting paste is termed vajralepa i.e. adamantine glue.[18] It is said, if this heated glue is used in the construction of temples, mansions, windows, walls, and wells as well as in fixing Murtis, it will last for Millions of years.[19]

Another glue of excellent quality is also used for the same purpose. It is composed of lac, Kunduru (secretion of deodar), Guggulu, soot, wood-apple (Feronia elephantum), Bilva kernel, fruits of naga (canthium parviflorum), Neem (Azadirachta indica), Tinduka (Diospyros paniculata), and Madana (randia dumetorum), madhūka (Cynometra ramiflora), Manjistha (Rubia cordifolia), resin, myrrh, and amalaka (emblica officinalis).[20]

---

[18] आमं तिन्दुकमामं कपित्थकं पुष्पमपि च शाल्मल्या: ।
बीजानि शल्लकीनां धन्वनवल्को वचा चेति ॥
एतै: सलितद्रोण: क्वाथयितव्योऽष्टभागशेषश्च ।
अवतार्योऽस्यच कल्को द्रव्यैरेतै: समनुयोज्य: ॥
श्रीवासक-रस-गुग्गुलु-भल्लातक-कुन्दुरुक-सर्जरसै: ।
अतसी. बिल्वैश्च युत:कल्कोऽयं वज्रलेपाख्य: ॥ (बृहत्संहिता, 56.1-3)
āmaṁ tindukamāmaṁ kapitthakaṁ puṣpamapi cha śālmalyāḥ ।
bījāni śallakīnāṁ dhanvanavalko vachā cheti ॥
ētaiḥ salitadroṇaḥ kvāthayitavyo'ṣṭabhāgaśeṣaścha ।
avatāryo'syacha kalko dravyairetaiḥ samanuyojyaḥ ॥
śrīvāsaka-rasa-guggulu-bhallātaka-kunduruka-sarjarasaiḥ ।
atasī. bilvaiścha yutaḥkalko'yam vajralepākhyaḥ ॥

[19] प्रसादहर्म्यवलभीलिंगप्रतिमासे कुड्यकूपेषु ।
सन्तप्तो दातव्यो वर्षसहस्रायुतस्थायी ॥ (बृहत्संहिता, 56.4)
prasādaharmyavalabhīliṁgapratimāse kuḍyakūpeṣu ।
santapto dātavyo varṣasahasrāyutasthāyī ॥ (Bṛhatsaṁhitā, 56.4)

[20] लाक्षा-कुन्दुरु-गुग्गुलु-गृहधूम-कपित्थ-बिल्वमध्यानि ।
नागफल-निम्ब-तिन्दुक-मदनफल-मधूक-मंजिष्ठा: ॥

There is a third type of glue named vajratala (literally meaning adamantine surface) which is constituted by the horns of cows, buffaloes, and goats, hair of donkeys, buffalo hide, cowhide, neem fruits, wood apples, and myrrh.[21] This mixture should also be boiled and reduced as before.

According to Prof. A.R. Vasudeva (2005: 54), even now, such glue is manufactured on a large scale after hydrolyzing the protein present in the starting raw materials such as horns, hooves, tendons, leather, and bone marrow. Gelatin of different grades, including edible grades, is also isolated from this mixture. In addition to these raw materials of animal origin, some plant materials were also employed to prepare the glues. Their gluten is separated from the starchy component, which helps prepare the cement.

One more variety of glue taught by Maya, known as vajrasaṅghāt (adamantine compound), was composed of eight parts of lead, two of bell metal, and one of iron

---

सर्जरस-रस-आमलकानि चेति कलकः कृतो द्वितीयोऽयम् ।

वज्राख्यः प्रथमगुणैरयमपि तेष्वेव कार्येषु ॥ (बृहत्संहिता, 56.5-7)

lākṣā-kunduru-guggulu-gṛhadhūma-kapittha-bilvamadhyāni /

nāgaphala-nimba-tinduka-madanaphala-madhūka-maṁjiṣṭhāḥ //

sarjarasa-rasa-āmalakāni cheti kalakaḥ kṛto dvitīyo'yam /

vajrākhyaḥ prathamaguṇairayamapi teṣveva kāryeṣu //

(*Bṛhatsaṁhitā, 56.5-7*)

21   गोमहिषाजविषाणैः खुररोम्णामहिषचर्मगव्यैश्च ।

निम्बकपित्थरसैः सह वज्रतलो नाम कल्कोऽन्यः ॥ (बृहत्संहिता, 56.8)

gomahiṣājaviṣāṇaiḥ khuraromṇāmahiṣacharmagavyaiścha /

nimbakapittharasaiḥ saha vajratalo nāma kalko'nyaḥ //

(*Bṛhatsaṁhitā, 56.8*)

rust.[22]

Another rock like a substantial variety of mortar taught by Maya is as under:

There should be five parts extract of beans, 9 and 8 parts jaggery and curd, respectively. Clarified butter (ghee) 2 parts, seven parts milk and hide six parts. There should be ten parts of Triphalā, coconut two parts, honey one part, and plantain three parts. In the paste thus obtained, 1/10th lime should be added. A large quantity than others of jaggery, curd, and milk is best. Add karka, honey, clarified butter, plantain, coconut, and bean in two parts of lime. Add water, milk, curd, Triphala, and jaggery when dry. The one part of the powder/paste thus obtained can grow 100 times. This compound is said as rock-like by the experts who know this technology.[23]

---

[22] अष्टौ सीसक भागाः कांसस्य द्वौ तु रीतिकाभागः।
मयकथितो योगोऽयं विज्ञेयो वज्रसंघातः॥

aṣṭau sīsaka bhāgāḥ kāṁsasya dvau tu rītikābhāgaḥ।

mayakathito yogo'yaṁ vijñeyo vajrasaṁghātaḥ॥

[23] पंचांशं माषयूषं स्यान्नवाष्टांश गुडं दधि।
आज्यं द्वयंशं तु सप्तांशं क्षीर चर्म षडंशकम्॥
त्रैफलं दशभागं स्यान्नालिकेरं युगांशकम्।
क्षौद्रमेकांशकं यंशं कदलीफलमिष्यते॥
लब्धे चूर्णे दशांशे तु युंजीतव्यं सुबन्धनम्।
सर्वेषामधिकं शस्तं गुडं च दधि दुग्धकम्॥
चूर्णद्वयंशं करालं मधुघृतकदलीनालिकेरं च माषम्।
शुक्तेस्तोयं च दुग्धं दधिगुडसहितं त्रैफलं तत् क्रमेण॥
लब्धे चूर्णे शतांशेऽशकमिदमधुना चानुवृद्धिं प्रकुर्या-
एतद् बन्धं दृषत्सादृशमिति कथितं तन्त्रविद्भिर्मुनीन्द्रैः॥

paṁchāśaṁ māṣayūṣaṁ syānnavāṣṭāṁśa guḍaṁ dadhi।

Till 250 years ago, the best quality mortar was manufactured in Chennai. One English Officer, Isaac Pyke, reported the method of making this mortar in Philosophical Transactions Vol. XXXVII in 1732, which is cited as under:

"Take fifteen fresh, pit-sand bushes well sifted; add fifteen bushels of stone lime. Let it be moistened or slacked with water in a typical manner and lay two or three days together.

Then dissolve 20 lb of jaggery, which is coarse sugar (or thick molasses), in water, sprinkle this liquor over the mortar, beat it up until all be well mixed and incorporated, and then it lies by in a heap.

Then boil a peck of a gram (a sort of grain like a tare, or between that and a pea) to a jelly, strain it off through a course canvass, and preserve the liquor that comes from it.

Take also a peck of myrobalans and boil them likewise to a jelly, preserving that water also as the other; and if you have a vessel large enough, you may put these

---

ājyaṁ dvayaṁśaṁ tu saptāṁśaṁ kṣīra charma ṣaḍaṁśakam ॥
traiphalaṁ daśabhāgaṁ syānnālikeraṁ yugāṁśakam ।
kṣaudramekāṁśakaṁ yaṁśaṁ kadalīphalamiṣyate ॥
labdhe chūrṇe daśāṁśe tu yuṁjītavyaṁ subandhanam ।
sarveṣāmadhikaṁ śastaṁ guḍaṁ cha dadhi dugdhakam ॥
chūrṇadvayaṁśaṁ karālaṁ madhughṛtakadalīnālikeraṁ cha māṣam ।
śuktestoyaṁ cha dugdhaṁ dadhiguḍasahitaṁ traiphalaṁ tat krameṇa ॥
labdhe chūrṇe śatāṁśe'ṁśakamidamadhunā chānuvṛddhiṁ prakuryā.
ētad bandhaṁ dṛṣatsādṛśamiti kathitaṁ tantravidbhirmunīndraiḥ ॥
¼Maya-Matam, Kāla-mūla-śāstram by Kapila Vātsyāyana)

three waters together; that is the jaggery water, the gram water, and the myrobalan. The Indians usually put a small quality of the lime therein to keep their labourers from drinking it.

The mortar beat up and, when too dry, sprinkled with this liquor proves extraordinarily good for laying brick or stone in addition to that. If this liquor proves too thick, dilute it with fresh water.

The mortar was also improved in this manner:

"Take course tow and twist it loosely into bands as thick as a man's finger (in England, Ox hair is used instead of this tow), then cut it into pieces of about an inch long and untwist it to lie loose; then sprinkle it lightly over the other mortar which is at the same time to be kept turning over, and so this stuff to be beaten into it, keeping labourers continually beating in a trough and mixing it till it be well incorporated with all the parts of the mortar, and whereas it will be subject to dry very fast, it must be frequently softened with the liquor made of jaggery, gram and myrobalans and some fresh water; and when it is so moistened and beat it will mix well, and with this, they build (though it is not usual to build common house-walls thus) when the work is intended to be very strong, as, Madras Church Steeple,

that was building when I was last there; and also for some ornaments, as columns, good arched work, or imagery set up in Gardens, it is this made."

Such reports show the quality of mortar manufactured in ancient India, which needs to be studied carefully by the experts.

Assembling Pillars

Pillars also played a significant role in the materialisation of constructions. Often constructions of big halls were based upon the posts. As such, assembling pillars also assumed a central part of civil engineering. Ancient Indian civil engineers invented five methods for mortise (a hole to receive a tenon) and tenon (the projection at the end of a piece of wood etc.) assemblies suitable for pillars. These were known as

1. Meṣayuddha

2. Trikhaṇḍa

3. Saubhadra

4. Ardhapāṇi and

5. Mahāvṛtta[24]

1. Meṣayuddha: When there is a central tenon (with a width), a third (that of the pillar), and a length twice or two and half times its width, this is Meṣayuddha (mortise and tenon) assembly).[25]

2. Trikhaṇḍa: Trikhaṇḍa assembly takes place with

---

24 मेषयुद्धं त्रिखंडं च सौभद्रं चार्थपाणिकम् ।
महावृत्तं च पंचैते स्तम्भानां सन्धयः स्मृता ॥
meṣayuddhaṁ trikhaṁdaṁ cha saubhadraṁ chārthapāṇikam /
mahāvṛttaṁ cha paṁchaite stambhānāṁ sandhayaḥ smṛtā //
        Maya-Matam, Kāla-mūla-śāstram by Kapila Vātsyāyana, śloka, 29

25 स्वव्यासकर्णमध्यर्धद्विगुणं वा तदायतम् ।
यशेकं मध्यमशिखं मेषयुद्धं प्रकीर्तितम् ॥ (*ibid. śloka* 30)
svavyāsakarṇammadhyardhadviguṇaṁ vā tadāyatam /
yaśekaṁ madhyamaśikhaṁ meṣayuddhaṁ prakīrtitam //

three mortises and three tenons arranged as a Svastika.[26]

3. Saubhadram: Saubhadram assembly comprises four peripheral tenons.[27]

4. Ardhapāṇi: An assembly is called ardhpāṇi (scarf joint) when half the lower and half the upper pieces are cut to size according to the thickness chosen for the pillar.[28]

5. Mahāvṛtta assembly: When a semi-circular tenon is at the centre, the assembly is called Mahāvṛtta. This is used for circular section pillars.[29]

Some precautions have also been laid down about assembly. It is said that assembling different parts of a pillar should be done below the middle, and any assembling done above will cause an accident; however, the assembly which brings together the bell capital and abacus gives the certainty of success. When a stone pillar, with its decoration, is to be assembled, this should be done according to the specific case.[30]

---

26   स्वस्त्यकारं त्रिखण्डं स्यात् सत्रिचूलि त्रिखण्डकम् । *(ibid. śloka, 31)*
     svastyakāraṁ trikhaṇḍaṁ syāt satrichūli trikhaṇḍakam /

27   पार्श्वे चतुःशिखोपेतं सौभद्रमिति संज्ञितम् ॥ *(ibid. śloka, 31)*
     pārśve chatuḥśikhopetaṁ saubhadramiti saṁjñitam //

28   अर्धं छित्वा तु मूलेऽग्रे चान्योन्याभिनिवेशनात्
     अर्धपाणिरिति प्रोक्तो गुहीतघ्नमानतः ॥ *(ibid. śloka, 32)*
     ardhaṁ chhitvā tu mūle'gre chānyonyābhiniveśanāt
     ardhapāṇiriti prokto guhītaghnamānataḥ //

29   अर्धवृत्तशिखं मध्ये तन्महावृत्तमुच्यते ।
     वृत्ताकृतिषु पादेषु प्रयुंजीत विचक्षणः ॥ *(ibid. śloka, 33)*
     ardhavṛttaśikhaṁ maghye tanmahāvṛttamuchyate /
     vṛttākṛtiṣu pādeṣu prayuṁjīta vichakṣaṇaḥ //

30   स्तम्भानां स्तम्भदैध्यार्धादधः सन्धानमाचरेत् ।

It may also be known that assembling of the vertical pieces is done according to the disposition of the different parts of the tree; if the bottom is above and the top is below, all chances of success are lost.[31]

स्तम्भध्योर्ध्वसन्धिश्छेद् विपदामास्पदं सदा ॥
कुम्भभाण्ड्यादिसंयुक्तं सन्धानं समपदां पदम् ।
सालंकारे शिलास्तम्भे यथायोगं तथाचरेत् ॥ (पइपक). *śloka*, 34-35)
stambhānāṁ stambhadairdhyārdhādadhaḥ sandhānamācharet ।
stambhadhyordhvasandhiśched vipadāmāspadaṁ sadā ॥
kumbhabhāṇḍyādisaṁyuktaṁ sandhānaṁ samapadāṁ padam ।
sālaṁkāre śilāstambhe yathāyogaṁ tathācharet ॥

31   स्थितस्य पादपस्यांगप्रवृत्तिवशतो विदुः ।
ऊर्ध्वमूलमधश्चाग्रं सर्वसम्पद्विनाशनम् ॥ (पइपक). *śloka*, 36)
sthitasya pādapasyāṁgapravṛttivaśato viduḥ ।
ūrdhvamūlamadhaśchāgraṁ sarvasampadvināśanam ॥

5

Temple Architecture

Temple architecture also developed on an enormous scale in India. Stupas and temples scattered all over India are a living testimony to this fact. Mayamatam, Mānasāra, Purāṇas, Samrāṅgaṇa Sūtradhāra, and Bṛhat Saṅhitā are some of the books that deal with this aspect of civil engineering. While constructing temples, emphasis was laid on the following issues:

a). Selection of stones to be used in temple construction and murti making

b). Tools used in the construction and carving of murtis

c). Material for murtis

d). Stone softening mixtures

Material for Murtis

While dealing with this aspect of civil engineering, the experts dealt with six types of stones available prominently in this country. They were

1. Hiraṇya rekhikā (of golden lining)

2. Samavarṇā (of uniform colour)

3. Tāmra (copper coloured)

4. Dhātu puṭita (mineral layered)

5. Vajra labdha (diamond-like)

6. Saiktālikā (Sand stone)32

The crooked stones, even with golden lining, were considered unfit for the murtis, as they caused disfigurement; their sight created a double impression. They were often considered purposeless.33

When a uniform coloured stone is found in one single colour, be it black, dark brown, or turmeric yellow, that stone should be taken for the Principal murti.34

A tough stone stone of copper-red colour was not considered fit for a murti.35

A stone unsevered from a rock was considered useable for sculpture halls, otherwise (severed) not.36

A stone of Dhātupuṭita type with smoke colour or

---

32 शिला तद्देशे लोकज्ञाने षड्ढा प्रथिता, हिरण्यरेखिका ।
समवर्णा, ताम्रा, धातुपुटिता, वज्रलब्धा, सैकतालिकेति ॥
śilā taddeśe lokajñāne ṣaḍḍhā prathitā, hiraṇyarekhikā /
samavarṇā, tāmrā, dhātupuṭitā, vajralabdhā, saikatāliketi //

(Vāstu Sūtra Upaniṣad, śail bhedanam, Sūtra, 9)

33 यदा हिरण्यरेखा कुटिला सोऽवलक्षणः ।
सा प्रतिमार्थहीना विरूपं प्रददाति । तद्दर्शनेन विकल्पभावः प्रभवति ।
दर्शनादेता अर्थहीना उच्यन्ते, वदन्ति, भवन्ति ।
एतदर्थं प्रतिमावैलक्षण्यं भवति ॥ (पइपकण्द्ध)
yadā hiraṇyarekhā kuṭilā so'valakṣaṇaḥ /
sā pratimārthahīnā virūpaṁ pradadāti /
taddarśanena vikalpabhāvaḥ prabhavati /
darśanādetā arthahīnā uchyante, vadanti, bhavanti /
ētadarthaṁ pratimāvailakṣaṇyaṁ bhavati // (paipakaṇddha)

34 समवर्णा यदा पूर्णशैलमेकवर्णं कृष्णं कृष्णपिङ्गलं
हारिद्रं वा मुख्यप्रतिमार्थं ध्येयं सा समवर्णा ।(पइपकण्द्ध)

35 कठिनतमगाढपिंगल शिला न ध्येया । (पइपकण्द्ध)

36 यदा शैलं भित्त्याकृतेरपृथक् प्रतिमाशालार्थं तद् ध्येयम् । (पइपक)

with pores was considered awful for murti-making.[37]

Bright yellow stones or that have in them or between their layers soft mineral lines were considered inferior from a construction point of view.[38]

Sandstones are often considered good for making yupa (pillars for yajña), altars, double pillars, murtis, and furnaces. The sculptors who possessed this knowledge were also greatly honoured by people.[39]

Tools used in the construction and carving of Murtis

Five types of chisels were made of steel, and each of two types, narrow and broad, were prescribed. The different types were known as

a). Lañjī (biting)

b). Lāṅglī (plough like)

---

[37] शिला धातुपुटिता धुम्रवर्णा व्रणोपेता रूपार्थं घोरेति। (पइपकण्द्ध)
śilā dhātupuṭitā dhumravarṇā vraṇopetā rūpārtham ghoreti ।
(paipakanddha)

[38] शैलान्तरे स्तराभ्यन्तरे कोमलरसबद्धा रेखाः सन्ति।
याः पीतप्रभाः सदाधमा भवन्ति श्रेष्ठैः शिल्पकारैः॥ (पइपकण्द्ध)
śailāntare starābhyantare komalarasabaddhā rekhāḥ santi ।
yāḥ pītaprabhāḥ sadādhamā bhavanti śreṣṭhaiḥ śilpakāraiḥ ॥
(paipakanddha)

[39] सैकतालिकशिला यूपयोनिमिथुनस्तम्भ वृषस्तम्भ कुण्डार्थं
विशिष्टा सा रूपार्थं ग्राह्या, तज्ज्ञानलब्धस्थापकाः।
समुपतिष्ठन्ति जनलोके सम्पन्ना महीयन्ते॥ (पइपकण्द्ध)
saikatālikaśilā yūpayonimithunastambha vṛṣastambha
kuṇḍārtham
viśiṣṭā sā rūpārtham grāhyā, tajjñānalabdhasthāpakāḥ ।
samupatiṣṭhanti janaloke sampannā mahīyante ॥ (paipakanddha)

c). Gṛdhradantī (like beak of vulture)

d). Sêcīmukhā (needle tipped)

e). Vajrā (made of diamond)

The chisels were beaten on the long mallet (a small wooden hammer) with a short mallet (a hammer made of steel). These chisels were used for breaking stones. All instruments were sharpened, dipped in cow's urine and then smeared with iṅgida (asafetida) oil and whetted in leather.[40]

Stone softening mixtures

Sculptors used to apply a stone softening mixture made of shell-solvent, kuṣtharasa (costus specious or arabicus. Note: in the forests of the Mahendra mountain, a tree is found whose bark is called stone-breaking, i.e. prastara-bhedī), sea salt, and the powder of the bark of

---

40 खनित्रपंचकं श्रेष्ठम् ।

भेदास्तु लांजी, लांगली, गृध्रदन्ती, सूचीमूखा वज्रा इति, सर्वे आयसा द्विविधा भवन्ति क्षीणाः
प्रशस्ताश्च, मुषलधरुभे मुषलदण्डेन खनित्रं घातयन्ति ।

प्रयोजयन्ति शिलाभेदेन तत् । सर्वास्त्राणि तीक्ष्णानि गवाम्बुपुटितानि ।

ततः इगिंडालेपितानि चर्मशाणितानि च ॥

(Vāstusūtra Upaniṣad, Sūtra 6, Śaila-bhedanam)

khanitrapaṁchakaṁ śreṣṭham /

bhedāstu lāṁjī, lāṁgalī, gṛdhradantī, sūchīmūkhā vajrā iti, sarve āyasā dvividhā bhavanti kṣīṇāḥ

praśastāścha, muṣaladharubhe muṣaladaṇḍena khanitraṁ ghātayanti /

prayojayanti śilābhedena tat / sarvāstrāṇi tīkṣaṇāni gavāmbupuṭitāni /

tataḥ igiṁḍālepitāni charmaśāṇitāni cha //

(Vāstusūtra Upaniṣad, Sūtra 6, Śila-bhedanam)

ukatsa tree, so that they may efficiently work on stones.[41]

---

41   शिल्पकाराः प्रलेपयन्ति द्रावकरसम् ।
शंखद्राव-कुष्ठरस-सैन्धवखर्पर-उक्तसवल्कलचूर्णेन सहितं शिलाद्रवणार्थमेवं रसचतुष्ट्यम् अनेन
मन्त्रेण दशाहमर्दनान्ते वैताने खनित्रं प्रयोजयन्ति खननमाचरन्ति भद्रेण शिल्पकाराः ॥
śilpakārāḥ pralepayanti drāvakarasam /
śaṁkhadrāva-kuṣṭharasa-saindhavakharpara-uktsavalkalachūrṇena
sahitaṁ śilādravaṇārthamevaṁ rasachatuṣṭyam anena mantreṇa
daśāhamardanānte vaitāne khanitraṁ prayojayanti
khananamācharanti bhadreṇa śilpakārāḥ //
                    (Vāstusūtra Upaniṣad, Sūtra 8, Śila-bhedanam)

6

Town Planning

Town planning expertise existed since Vedas's time in this country. Towns were known as puras in the Vedic and post-Vedic periods. Ancient Indians were well versed in planning a township, constructing roads in towns and mountainous terrain, planning the course of roads, and regulating the construction of houses-like mandatory spacing between houses. There are examples of each of these in Vedic and post-Vedic Sanskrit literature.

Ancient Law for housing

A.L. Basham (2003:16)[42] observes that the houses had bathrooms, the design of which shows that the Harappan, like modern India, preferred to take his bath standing by pouring pitchers of water over his head. The bathrooms had drains, which flowed to sewers under the main streets, leading to soak pits. The sewers were covered throughout their length by large brick slabs. The unique sewerage system of the Indus people must have been maintained by some municipal organizations and is one of the most impressive of their achievements. No other ancient civilization until that of the Romans had so efficient a system of drains.

We came across that some laws, not prescriptive but descriptive, were formulated in the Maurayan period

---

42   The wonder that was India, 3rd Edition.

regarding the construction and spacing of houses. Accordingly, the distance between two houses was three steps; the rule applies to houses with Chajjas or Varandas. The distance between the roofs of two adjacent houses was to be kept four Angulas, or they may be allowed to touch each other. A closeable window of one 'kiṣku' dimension was prescribed towards the lane. For light window was allowed to be made at a higher level. Rest the onus was placed on house owners to do whatever was good for them to avoid bad.[43]

Construction of Roads in Plains

The surface of roads was prescribed to be made like the back of a tortoise, along with beautiful bridges and flyovers wherever required. The drains were prescribed to be dug in the sides of roads for drainage of water.[44] The administration also used to appoint unemployed

---

43   सर्ववास्तुकयोः प्रक्षिप्तयोर्वा शालयोः किष्कुरन्तरिका त्रिपदी वा ।

तयाश्चतुरंगुलं नीप्रान्तरः समारूढकं वा ।

किष्कुमात्रमणिद्वारमन्तरिकायां खण्ड फुल्लार्थमसम्पातं कारयेत् ।

प्रकाशार्थमल्पमूर्ध्ववातायनं कारयेत् । सम्भूयवा गृहस्वामिनो यथेष्टं कारयेयुरनिष्टं वारयेयुः ।

sarvavāstukayoḥ prakṣiptayorvā śālayoḥ kiṣkurantarikā tripadī vā /

tayāśchaturaṁgulaṁ nīprāntaraḥ samārūḍhakaṁ vā /

kiṣkumātramaṇidvāramantarikāyāṁ khaṇḍa

phullārthamasampātaṁ kārayet /

prakāśārthamalpamūrdhvavātāyanaṁ kārayet / sambhūyavā

gṛhasvāmino yatheṣṭaṁ kārayeyuraniṣṭaṁ vārayeyuḥ /

(*Arthaśāstram of Kuṭilya, 3.8.19*)

44   कूर्मपृष्ठा मार्गभूमः कार्यगारम्यैः सुसेतुकाः ।

कुर्यान्मार्गान् पार्श्वखातान् निगमार्थे जलस्य च ॥

kūrmapṛṣṭhā mārgabhūmaḥ kāryagāramyaiḥ susetukāḥ /

kuryānmārgān pārśvakhātān nigamārthe jalasya cha //

(*Śukranītiḥ, 1.165*)

people to repair broken roads every year with lime and gravel.[45]

Different roads had different dimensions. For instance, Roads used by cars were of 4 dandas' (24 feet) width. Royal paths, National highways, state highways, and local highways, roads leading to lakes, pastures, and cremation grounds were eight dandas (48 feet) each. Bridges and forest roads were of 4 dandas (24 feet) each. Roads for chariots 5 aratni (5 feet), roads for cattle, small animals, and humans were of 2 aratni (1 foot) each.[46]

Construction of Roads in Mountainous Terrain

The mountain roads were prescribed twice or thrice as broad as those in plains. Streams could have small or large bridges as per requirements.[47]

---

[45] मार्गान्सुधाशर्करैविघट्टितान् प्रतिवत्सरम् ।
अभियुक्ता निरुद्योगैः कुर्याद् ग्राम्यजनैर्नृपाः ॥ (Śukranītiḥ, 1.166)
mārgānsudhāśarkaraivighaṭṭitān prativatsaram /
abhiyuktā nirudyogaiḥ kuryād grāmyajanairnṛpāḥ //

(*Śukranītiḥ, 1.166*)

[46] चतुर्दण्डान्तरा रथ्याः । राजमार्ग-द्रोणमुख-स्थानीय-राष्ट्रविवीतपथाः संयानीय-व्यूह-श्मशान-
ग्राम-पथाश्चाष्टदण्डाः चतुर्दण्डः सेतुवनपथः । द्विदण्डो हस्तिक्षेत्रपथः । पंचरत्नयो रथपथश्चत्वारः
पशुपथो द्वौ क्षुद्रपशु-मनुष्यपथः ।
chaturdaṇḍāntarā rathyāḥ / rājamārga-droṇamukha-sthānīya-
rāṣṭravivītapathāḥ saṁyānīya-vyūha-śmaśāna-grāma-
pathāśchāṣṭadaṇḍāḥ chaturdaṇḍaḥ setuvanapathaḥ / dvidaṇḍo
hastikṣetrapathaḥ / paṁcharatnayo rathapathaśchatvāraḥ
paśupatho dvau kṣudrapaśu-manuṣyapathaḥ /

(*Arthaśāstram by Kauṭilya, 2.20.4*)

[47] प्रतोली मानगणना द्विगुणं त्रिगुणं तु वा ।
कुल्या वेतुयुतं वापि महासेयुतं क्वचित् ॥
pratolī mānagaṇanā dviguṇaṁ triguṇaṁ tu vā /

Draining of rainwater was allowed to pass from the sides of the culvert to a fast-flowing river nearby or located at a distant place or to a river or reservoir. The culvert was made high in the middle and low on the sides.[48]

The road on the culvert was made hard and compressed with the stones and brickbats.[49]

Svastika Road Plan

A Svastika road plan was laid out, according to which cities and villages were equipped with a Svastika road scheme.

Accordingly, six roads go east and north. The roads approaching these six are on the outside. The street layout was a road junction. The east and north roads

---

kulyā vetuyutaṁ vāpi mahāseuyutaṁ kvachit ǁ

(Vāstuśāstram, Mārgalakṣaṇam Adhyāya 68, śloka 4)

48   पक्षयोस्सलिलस्रावो घनकल्पनमेदुरः ।

महाजलगतिस्स्यादप्यादुरतरेऽपि ॥

नदीमेलनकल्पा वा चाब्धिमेलनकल्पनम् ।

मध्योन्नतिः पार्श्वेनिम्नं घनकत्पनम् ॥

pakṣayossalilasrāvo ghanakalpanameduraḥ ǀ

mahājalagatissyādapyāduratare'pi ǁ

nadīmelanakalpā vā chābdhimelanakalpanam ǀ

madhyonnatiḥ pārśvenimnaṁ ghanakatpanam ǁ

(Vāstuśāstram, Mārgalakṣaṇam Adhyāya 68, śloka 5-6)

49   शिलाखण्डैः कुम्भिखण्डैर्घनर्मेदुरकल्पनम् ।

गजपादैरश्वपादैर्मर्दितं च दृढिकृतम् ॥

śilākhaṇḍaiḥ kumbhikhaṇḍairghanarmedurakalpanam ǀ

gajapādairaśvapādairmarditaṁ cha dṛḍhikṛtam ǁ

(Vāstuśāstram, Mārgalakṣaṇam Adhyāya 68, śloka 7)

shall be only 4.[50]

Svastika was a scheme of parallel roads of cities, or, say, the layout of principal roads forming a svastika. The features of the svastika format were:

1. The central districts had six roads running North to south and six east to west.

2. In the periphery, there were only four roads in each direction.

3. In the periphery, all roads used to run in a clockwise direction.

4. The apparent benefits of such an arrangement were

5. The plan provided higher road density in the town's central part than the periphery, which would correspond with the traffic density in any city.

6. The roads permit one-way traffic in each direction.

7. The arrangement discourages traffic skirting the city from going through the central part of the city.

---

50 स्वस्तिकमुदितं ग्रामे यथा तथा स्वस्तिकं चैव ।+

प्रागुत्तरमुखमार्गाः षड्डभीष्टास्तु तद्बाह्ये ॥

प्रागिव मार्गोपेतं वीथिपदं स्वस्तिकं चैव ।

प्राचीनोदीचीनाश्चत्वारश्चैवमार्गाः स्यु ॥

prāguttaramukhamārgāḥ ṣaṭṣaḍabhīṣṭāstu tadbāhye ॥

prāgiva mārgopetaṁ vīthipadaṁ svastikaṁ chaiva ।

prāchīnodīchīnāśchatvāraśchaivamārgāḥ syu ॥

(Mayamatam, Kāla-Mūla-Śāstram, Kapil Vatsyāyanaḥ, śloka, 60-61)

7

Mechanical Engineering

As pointed out above, three types of technology developed in Ancient India - māntrika, tāntrika, and yāntrika. *Yāntrika* technology was the nomenclature given to mechanical engineering, and machines were known as yantras.

In Ancient India, the use of yāntrika technology was widespread. The word "yantra" is derived from the root *yam* to control and has been freely used in ancient India for any mechanical work.

Scientific publications like the *Yantrāraṇava* and the *Yantra-sarvasva* specially deal with the technique and operation of machines and mechanical devices.

Yantrasarvasva describes yantra, i.e. machine, as a system of generation of power/energy or motion through continuous movement or rotation of shafts, wheels, or wedges (gear).[51]

Various Yantra

Here we shall mention various yantras, from very simple ones to very complicated ones.

1. *Ghaṭi-yantra*: It was water-pulley, which was the

---

51 दंडैश्चक्रैश्च दन्तैश्च सरणि भ्रमणादिभिः ।
शक्तेरुत्पादनं किं वा चालनं यन्त्रमुच्यते ॥
*daṁḍaiśchakraiścha dantaiścha saraṇi bhramaṇādibhiḥ ǀ
śakterutpādanaṁ kiṁ vā chālanaṁ yantramuchyate ǁ*

most common yantra used in the well to pull water for drinking and irrigation in the fields

2. *Taila-yantra:* It was an oil-presser. Bhāgavat Purāṇa refers to an oil-expelling device describing a gear arrangement.

3. *Ikṣu-yantra:* It was a cane presser.

4. *Hala*: It was a plough.

5. *Charas*: It was a water lift.

6. **Stone grain-grinder:** In Tamil, the stone grain-grinder is called *"entiram."*

7. **Physical Balance:** The *Arthaśāstra* of Kauṭilya (2.19: 2.24; 2.35.16) describes making physical balance as under:

Make a balance at 72 finger width of an iron piece of 25 pala weight; bind a ring of 5 pala weight on it. Then on the opposite side, make a gradation of 11, 12, 15, and 20 palas. From there, make steps of 10 palas up to 100 palas. [52]

Note: One pala = 1400 gm/ 40 or 35gm

---

[52] पंचविंशतिपललोहां द्विसप्तत्यंगुलयामां समवृत्तां कारयेत् ।
तस्याः पंचपालिकं मण्डलं बध्वा समकरणं कारयेत् ।
paṁchaviṁśatipalalohāṁ dvisaptatyaṁgulayāmāṁ samavṛttāṁ kārayet I
tasyāḥ paṁchapālikaṁ maṇḍalaṁ badhvā samakaraṇaṁ kārayet I
ततः कर्षोत्तरपलं फलोत्तरदशपलं द्वादश पंचदश विंशतिरिति पदानि कारयेत् ।
अक्षेषु नद्धीपिनद्धं कारयेत्
tataḥ karṣottarapalaṁ phalottaradaśapalaṁ dvādaśa paṁchadaśa viṁśatiriti padāni kārayet I akṣeṣu naddhrīpinaddhaṁ kārayet

There is no dearth of references to machines and mechanical devices in ancient Indian literature, from Vedas to post-Vedic literature. For instance-

8. **Vimānas**: Vedas (Śvetavārāha Kalpa) talk of Vimānas. This issue will be dealt with separately in the chapter on aeronautics. *Bhṛgu Saṁhitā* classifies the vehicle flying in the sky as Agniyāna and the vehicle flying in space as Vyomyāna.[53]

9. **Naukā:** *Bhṛgu Saṁhitā* classifies the vehicle sailing in waters as Naukā (Ship). Here it may be known that boat is not the proper translation of the word Naukā.[54]

10. **Ratha**: One of the best creations of the most ancient architects of this country is the chariot (Ratha). The Ratha-kāra of Vedic times was ever a person of importance. *Bhṛgu Saṁhitā* classifies the operated on-road as Ratha. Here it may be known that chariot is not the proper translation of Ratha.[55]

11. **Chakra**: Wheels and their spokes and axle are frequently mentioned in the Vedas The wheel is the most significant fundamental invention of humankind. The wheel gave rise to everything from transportation to modern-day machinery. Without the invention of the wheel, no mechanical advancement was possible.

12. **Maththānī**: Mathānī was an apparatus used as a churning rod for churning milk, which is found

---

53   जलेनौकेव यानं स्याद् भूमियानं रथं स्मृतम् ।
आकाशे अग्नियानं च व्योमयानं तदेव च ॥
jalenaukeva yānaṁ syād bhūmiyānaṁ rathaṁ smṛtam |
ākāśe agniyānaṁ cha vyomayānaṁ tadeva cha ||

54   Op. cit.

55   Op. cit..

mentioned nowhere except India.

13. **Agnīddha Yantra**: It was used during the Vedic period for producing fire. This shows that since Vedic period itself Indians had the controlled use of fire. They used fire in Yajñas and cooking, as s ssource of light, for providing heat and warmth and it was also used by them to scare away wild animals.

14. **Sūtraveṣṭana**: It was a shuttle of Handloom used since Vedic times. It was called by various names as Nāḍichira, Nirveṣṭana, Pravāṇī, etc.

15. **Matsya yantra**: Mahābhārata mentions a Matsya-yantra or the revolving wheel with a fish that Arjuna had to shoot to win Draupadi in the svayaṁvara. This has a late echo in the Gadyachūḍāmaṇi and Kṣatra-chūḍāmaṇi of the Jain poet Vādibhāsiṁha of the 12th century, in a similar context of svayaṁvara, the test stipulated here being the piercing of three boars placed within a yantra called *Chandraka-yantra.*

16. **The magnetic compass**: The magnetic compass was first used in India for navigational purposes at sea and was known as 'Matsya yantra' (which roughly translates to fish machine). It used to be made of a metallic piece shaped into a fish and placed in a cup of oil.

16. **Indradhvaja:** *Adiparva* of the *Mahabharata* (64) describes the festival of Indra's banner inaugurated by Uparicara Vasu, which after that continued to be celebrated as a great national festival, by the Kings especially. The *Rāmāyaṇa,* which refers to Indra's Banner more than once, takes it as a *yantra-dhvaja.* The Flagstaff representing Indra was an elaborate affair, with many things hanging around it all around, and its raising and

bringing down after a 10-day festival was impressive sights. The falling of Bharata in *Ayodhyā* (77.9) is compared to the *yantra-dhvaja* of Indra falling. That the flag was raised and pulled down using fittings and ropes is known from another reference, in the *Yuddhakanda,* where Rama and Lakshmana lying down in Indrajit's *Nagapasa* are compared to the Indra banner after the ropes have been let loose. The full particulars of the erection and dismantling of the *Indradhvaja* are set forth by Bhoja (336-392 A.D.) in a chapter of over 200 verses exclusively devoted to it in his *Samarāṅgaṇa Sūtradhdra.* The central pole, the pedestal, the painted flag itself, the subsidiary fittings, the dolls to be hung on it, its outstretched arms, the six ropes attached to it, and *several yantras* fitted to it for raising it and bringing it down are described by Bhoja (336-392 A.D.).

17. **Asma-yantra**: In the *Harivamsa,* a supplement to the *Mahabharata* (3101 BC), there is mention of the stone-throwing machine, *Asma-yantra,* in the battle with Jarasandha (2. 42. 21). Kauṭilya's Arthśāstra (15th century B.C.) describes 32 machines including one for throwing stones and a weighing balance of unequal beam, indicating theoretical development of mechanics and sophistication of technology to sustain it.

18. **Perpetual motion machine**: Aryabhaṭa (2766 BC) mentions a perpetual motion machine.

19. **Globe**: Āryabhaṭīyam (Golapādaḥ, 2) mentions a globe (earth model) made of a small wooden sphere with its weight distributed uniformly on all sections fixed on its axis should be rotated by mercury, oil, and water for

one revolution per day.[56]

Note: Here, a globe is mentioned, which is kept on its axis, passing through the north and south poles and is free to rotate on bearings. Assuming that the first zodiac Aries is on the eastern horizon, a nail is inserted on the western side on its equator. Āryabhaṭa describes this point as a centre of Virgo and Libra. One end of the thread is hooked to this nail, the thread is wound anti-clockwise, and a pumpkin is suspended at the other end. A pot with a circular cross-section and deep bottom (*sama vṛtta* and *dīrgha tala*) is placed between North and South towards the east. It is filled with water having a volume equivalent to six ghatikas. Here it means that the volume of water should be such that it empties from the pot through a small hole made at the bottom in a period of six ghatikas. The water is filled, and the float is placed on the water's surface at sunset. Then as the water comes out through the hole, its level should fall, and the float should descend, making the sphere rotate one revolution per day automatically.

20. Carts: *Vāstuśāstram* describes the norms of an ideal cart in the chapter of Ratha-lakṣaṇa as under:

The outside length of the line between hubs of the two wheels may be equal to six (108 inches or 9 feet), seven (126 inches or 10¼ feet) or eight (144 inches or 12

---

56 काष्ठमयं समवृत्तं समन्ततः समगुरुं लधुं गोलम् ।

पारद-तैल-जलैस्तं भ्रमेत् स्वधिया च कालसमम् ।

kāṣṭhamayaṁ samavṛttaṁ samantataḥ samaguruṁ ladhuṁ golam /

pārada-taila-jalaistaṁ bhramet svadhiyā cha kālasamam /

feet) sets of two vitastis (i.e.18 inches or 1¼ feet).[57]

Weight should be balanced in the middle. The centre of the laden weight may be stretched forward. The laden weight may be placed between the two axis.[58]

It is always ideal to keep the weight in the cart's centre; the direction for error is to move the weight to the front. This is essential for any animal-vehicle since the locomotion is in the front. If the weight were to be towards the rear, then there would be the risk of the vehicle tilting. This format answers the norm of prescribing the ideal and suggests the direction for error.

Note: One *vitasti* is the distance between the prolonged thumb and a little finger or between the wrist and the tip of the finger and is said to be equal to 12 angulas or about 9 inches.

In Upala-yantras more interesting references are made by Valmiki (24th Tretāyuga, i.e. 18 million years ago) to yantras in the field of battle; the continuity of this tradition we see later in the *Arthaśāstra* of Kautilya (15th century BC). The fortifications include equipment in the form of yantras. In *Ayodhyakanda* (100.53), in

---

57 द्विचक्रबाह्ये विस्तारं षट् सप्ताष्टवितस्तियुगम् ।
चक्रनाभिच्छायामोक्षान्तरस्य च ॥

dvichakrabāhye vistāram ṣaṭ saptāṣṭavitastiyugam /
chakranābhichchhāyāmokṣāntarasya cha //

58 मध्यभारोपरि तुला मध्यनिर्गमनाग्रतः ।
अक्षमोक्षान्तरं चक्रमद्भारोपयानकम् ॥
पेतिकाकारसंयुक्तम् अयःपट्टैर्दृढीकृतम् ।

madhyabhāropari tulā madhyanirgamanāgrataḥ /
akṣamokṣāntaram chakramadbhāropayānakam //
petikākārasamyuktam ayaḥpaṭṭairdṛḍhīkṛtam /

the *Kaccit-sarga,* while enquiring about measures of defence, Rama asks Bharata whether the fort is equipped with yantras. As a city built by Maya, Lanka is naturally equipped and full of yantras. The city personified as a lady, is called *yantra-agara-stani,* informing us of a special chamber filled with yantras. *(Sundara* 3. 18). In his account to Rama of the fortifications of Lanka, Hanuman says in *Yuddha* 3.12 that Lanka has four big gates and that each gate is furnished with solid and huge yantras that can hurl mortars and missiles (*Upala-yantras):*

तत्रेषु उपलयन्त्राणि बलवन्ति महान्ति च

tatreṣu upalayantrāṇi balavanti mahānti cha

Moreover, over the moats are bridges controlled by numerous big yantras (3.16, 17). That such yantras were employed on the field is seen in a description of Kumbhakarna, in *Yuddha* 61.32, where his giant figure striding the streets of Lanka is compared to a vast yantra that has been set up.

उच्यन्तां वानराः सर्वे यन्त्रमेतत् समुच्छ्रितम्

uchyantāṁ vānarāḥ sarve yantrametat samuchchhritam

Mammoth machines turned by many persons and making a terrific noise are to be seen in a description of Ravana, soon after he got the news of Indrajit's death. As he gnashed his teeth in fury, the noise was heard of a big catapult being turned by Danavas.

दन्तान् विदशतस्तस्य श्रूयते दशनस्वनः ।
यन्त्रस्यावेष्ट्यमानस्य महतो दानवैरिव ॥ 6.93.23

dantān vidaśatastasya śrūyate daśanasvanaḥ |
yantrasyāveṣṭyamānasya mahato dānavairiva ॥ 6.93.23

The mention here of Danavas in connection with

machines must be noted. Later we see references to machines, associating the Yavanas, especially with them. The reference gains some significance when we consider the relationship of Maya with the Asuras and of the Asuras with Iran and the near West and the continuous contact that ancient India had with these neighbouring civilizations on the West.

The *Arthasastra* of Kautilya (15th century B.C.) is one of the books of culture which throw a flood of light on the particular epochs in which they arose. This work of 1500 B.C. is a treatise on statecraft and speaks of yantras in connection mainly with battles but also with architecture to some extent. An early work, a theoretical treatise, and a text of excellent reputation, the *Arthasastra* forms our most valuable document on the subject of yantras. Before we come to its account of the main yantras of warfare, we shall note some of the other machines met within Kautilya's work.

21. *Yantra-yukata-sopāna:* In his *Arthaśāstra* (2. 5), Kautilya (15th century B.C) refers to a dug-out *Bhumi griha* and mentions for it a machine-operated staircase which can be thrown in and withdrawn (*Yantra-yukata-sopāna).*

22. *Yantra-torana* (a machine-operated arch): It may be recalled here that in the political play of Viśakhadatta in which the author of the *Arthasastra* is the leading character, Kautilya is made to use this device against Vairocaka, brother of Parvatakesvara, the unwanted partner; Kautilya gets advance intelligence of the ruse employed by Rākṣasa through an architect named Daruvarman; the architect had erected a *yantra-torana,* a machine operated arch, which could be brought down by the drawing out of one of the fastening rods; this had been set up to kill Chandragupta as he entered the palace for the coronation, and Kautilya offers Vairocaka as a victim to this *yantra-torana*. This may be compared to the *Visvāsa-ghāti,* to be noticed below.

23. Bed chamber: A second ruse mentioned by Kautilya for the same purpose is a bed chamber in which part of the flooring has a mechanical contrivance; underneath this part of the floor is a deep cavity or a pit with pikes, and over it is placed the bed of the unwanted person; after he goes to sleep, the flooring is released and down goes the poor man with his bed.

Machines used in Military and warfare: Several types of yantras were also used for actual warfare, and forts were equipped with yantras. Yantras on the field are said to be attended to by specially trained persons. Kautilya devotes a particular chapter to the armoury. Ayudhāgara (2.18) is the main section speaking of military yantras. Kauṭilya divides the yantras into stationary and mobile - *Sthira* and *Cala yantras*. The former class comprises:

24 *Sarvatobhadra*: It was a sharp-edged wheel placed on a wall and rotated to launch explosive shells; according to others, it is also called *Siddhabhūmirika-yantra*.

25. *Jamadagnīya*: It was a fire-arm.

26. *Bahumukha*: This is an elevation and a mount for archers; it is leather covered and is as high as the wall to enable archers to shoot all around.

27. *Visvāsaghāti*: An iron bar across the path in the approaches to the city, which, operated by a machine, falls and pounds a man. It belongs to the class mentioned above for killing unwanted persons and the *yantra-torana* mentioned in the *Mudrarārākṣasa*.

28. *Sanghaṭi:* It used to set fire to enemy fortifications. It is called an *Agni-yantra.*

29. *Yānaka* **or** *Yānika* is a yantra moved on wheels; it discharges explosive shells.

30. *Parjanyaka* is an Udaka-yantra, a fire-quencher.

31. *Bahus* are two arm-like pillars which, when released from either side by a yantra, press to death a person between them; this appears to be an instrument of torture.

32. *Urdhvabāhu* is similarly an overhead column that comes down upon a man and puts him to death.

33. *Ardhabāhu* is the same as Bahu but is of a smaller size.

The *Chala-yantras*

34. Automatic Gun: While detailing the exact methods to be adopted for finishing off enemies and unwanted persons (12. 5), Kautilya speaks of machines that could be conveniently pressed into service. When the unwanted person is entering a temple, from an overhead yantra there could be released on his head a shot to kill him instantly.

35. *Hasti-yantras:* Kauṭilya (15th century B.C.) mentions a machine called Hasti-yantra intended to smite the elephant herds. These were robotic machines behaving like elephants in general or specially designed to scare elephants or enemies. In 13.4, Kauṭilya (15th century B.C.) advocates using yantras to devastate an enemy place full of defence erections.

36. *Pāñcālika*: Its use is outside the fort walls, in the moat; amid the water, its sharp protruding points prevent enemies' progress.

37. *Devadaṇḍas* are long cylindrical cannon-like things placed on parapet walls. Bhattasvamin gives them another name also, *Pratitaroca*.

38. *Śukarikā* is a vast robotic machine shaped like a pig or bellows, placed on the path as an obstruction and as a buffer to stop shells launched by the enemy.

According to some others, Śukarikās are to prevent enemies from quickly getting up the ramparts; on this view, they were probably closely suspended all along the walls to prevent the enemy soldier from getting a foothold.

39. *Musala and Yaṣṭi* -are well-known. Hastivāraka was a machine to drive away elephants. Dandin's *Avantisundari* mentions a machine that hurls-heavy iron rods to smite and demoralize the elephants. Hastiparigha was a machine to catch the enemy's elephants.

40. *Tālavṛnta*: A Tālavṛnta mentioned is explained as a Vātacakra. It was a device to create a tempest that could demoralize the enemy ranks. The observation of Philostratus relating to Alexander's invasion of India is that Indians drive the enemy off "using tempest and thunders, as if from heaven." It was like the *Vāyavya-astra*."

After Mudgara, Gadā, Sphriktalā, a picked missile, and Kuddāla are mentioned in the following:

41. *Asphoṭima* has four feet, is covered by hide, has a projectile, and throws stones.

42. *Udghāṭima* was a machine that demolished walls with iron bars fitted.

43. *Utpāṭima* is interpreted as the *Syena-yantra,* which uproots and tears up things.

44. *Śataghni:* And before the Tri" ula and Cakra, which are known, there occurs the Sataghni, the killer of hundreds, which is mentioned in all descriptions of warfare in old literature. Bhattasvamin takes it as a vast, cannon-like, cylindrical thing with wheels placed on the parapet.

Bhattasvamin also quotes a verse on yantras in general as of three kinds, *Vyādhita, Bhrāmita,* and *Bhārayukta*: the first operates by being pressed, the second by rotation, and the third by its sheer weight.

व्याधितं भ्रामितं चैव भारयुक्तं च कारयेत् ।
पीडनाद् भ्रमनाद् भाराद् त्रिधा यन्त्रं प्रवर्तते ॥

vyādhitaṁ bhrāmitaṁ chaiva bhārayuktaṁ cha kārayet /
pīḍanād bhramanād bhārād tridhā yantraṁ pravartate //

The Jain Sutras, which may go back to Mauryan times (15th century B.C.) and attained their present form in the 6th century A.D., know these yantras. The *Bhagavali Sūtra* lists eight finalities, one of which is the last fight with huge stones thrown as missiles.

45. *Mahāśilā-kantaka:* The same Sūtra describes the war between Kūṇīya and Chedaga, which was called as Mahāśilā-kanaka-Saṅgrama, i.e. a battle in which the stone complex besets one. The text itself elaborates that such a heavy stone shower has been in operation that even if a blade of grass, a piece of wood, a leaf, or a pebble struck one, one got the fright of a stone.

The second name was Ratha-musala-Saṅgrāma, where an unmanned chariot moved without horses and worked carnage in the enemy lines.

Hoernle has the following note on these two, Mahāsilā and Ratha-musala, which are war machines: "The Mahāsilākantaka must have been some engine of a war of the nature of a catapult that threw big shells or stones. It created such a panic among the enemy that all fled the Ratha-Musala would seem to have been a sort of 'scythed chariot'[59] such as the ancient Persians used to employ in war but furnished with clubs instead of scythes. It would also seem to have been provided with some remote-controlled machinery to propel it, as it is described as having moved without horses or driver."

46. Firearms and gunpowder: The above marks off one period in the history of these yantras. One of the questions that naturally agitate one's mind at this stage is that of firearms in ancient India. Throwing arrows with fire or combustibles on them is undoubtedly very old and forms one of the things prohibited by Manu in righteous warfare. Some of the other arms and machines mentioned by Kautilya do appear to be firearms.

Writers like Elliot[60] and Oppert[61] have argued not only

---

[59] *Cf.* Roger Bacon (latter part of the 13th century A.D.) in his *Epistolo de secretis operibus:* "Also cars can be made so that without animals they will move with unbelievable rapidity; such we opine were the scythe-bearing chariots with which men of old fought." (THORNDIKE, *op. cit.,* Vol. I I , p. 654.)

[60] H. M. ELLIOT, History of India, Vol. VI, Appedix, pp. 455-482

the prevalence of firearms in ancient India but also that India being the original home of gunpowder, the fire meant being not merely naphtha[62].

We find the use of fire-arms as early as *Sukraniti* (8th century A.D) and the *Nitiprakāsikā* of Vaiśampāyana

47. Flying ladders, wooden oxen and rolling horses: Elliot also collected the following data about particular mechanical objects used in warfare in India: the *Majmalu-t-Tawarikh* says that the experts advised[63] their king to employ on the battlefield an elephant of clay which, as soon as the enemies were sufficiently near, would explode and blow them up; more than the mere explosion, the device for timing it is to be noted here. In a Chinese account of India, reference is made to the use in Indian warfare of flying ladders, wooden oxen, and rolling horses.

48. Aerial Vehicles: Aerial vehicles will be described in the chapter on Aeronautical Engineering.

49. Robotic Soldiers: Lalitālaya, who is the actual character figuring in the narrative, and is said to excel his father, is credited with the following achievements, the description of which forms a brief treatise on yantras. He

---

61   G. OPPERT, On the Weapons, Army Organisation and Political Maxims of the Ancient Hindus with special reference to Gunpowder and Firearms. 1880. See also I. H. Q Vols. VII & VIII.

62   On the early use of gunpowder in the East see also M. Akram Makhdoomee, Mechanical Artillery in Medieval India. *(Joumal of Indian History,* Vol. XV, p. 185)

63   Ibid., pp. 473, 476, 463. See also Makhdoomee, op.cit pp. 189-155

created robots (mechanical men) and arranged for the exhibition of a mock-duel between them. The robotic fighters are also mentioned by Bhoja (Chap. 31, see Appendix).

50. Artificial Cloud: Lalitālaya created an artificial cloud and brought down heavy showers. Kauṭilya (16th century B.C.) also mentions a yantra called Parjanyaka, which used to make the artificial rain.

51. War Machine: Lalitālaya also devised a war machine with yantras from which shafts as stout as pestles were discharged by him on the heads of elephants.

दृष्टेऽपि तस्मिन् विस्मयस्पृशो जनस्य 'अयं किल यन्त्रपुरुषैः द्वन्द्वैः युद्धमहीनादर्शितवान्। अनेन किल अलीकजलधरधाराजल- जालदन्तुरितमन्तरिक्षं कृतम्। एष किल यन्त्रमयमिन्द्रजालकं कृतवान्। एष किल संख्येषु असंख्यानां युगपदेव भिनत्ति शत्रुहस्तिनां मस्तकस्थलानिमुसलमात्राभिरिषुभिः। अमुना किल द्रविडभाषया शुद्रकचरितमुपनिबद्धम्। अस्यकिल पित्रा यवनानप्यतिशयानेन क्षुधितोऽयमितियन्त्रेणाभिधावितम्। अयं ततोऽप्यधिकः किल इत्येवमासन् विकसितकुतुहलाः प्रलापाः।

dṛṣṭe'pi tasmin vismayaspṛśo janasya 'ayaṁ kila yantrapuruṣaiḥ dvandvaiḥ yuddhamahīnādarśitavān / anena kila alīkajaladharadhārājala- jāladanturitamantarikṣaṁ kṛtam / ēṣa kila yantramayamindrajālakaṁ kṛtavān / ēṣa kila saṁkhyeṣu asaṁkhyānāṁ yugapadeva bhinatti śatruhastināṁ mastakasthalānimusalamātrābhiriṣubhiḥ / amunā kila draviḍabhāṣayā śudrakacharitamupanibaddham / asyakila pitrā yavanānapyatiśayānena kṣudhito'yamitiyantreṇābhidhāvitam / ayaṁ tato'pyadhikaḥ kila ityevamāsan vikasitakutuhalāḥ pralāpāḥ /

Six categories of Yantras: Lalitālaya is said to be the master of all kinds of yantras; the varieties mentioned in this connection are six, Sthita, Cara, Dhāra, Dvipa, Jvara (yantra used for biological warfare), and Vyāmiśra. Other texts speak of the classes of yantras as two and five, giving the classifications somewhat differently. The Sthita or Sthira and Cara-stationary and mobile-is a classification done by Kauṭilya, as already cited above. Dhāra is a kind of yantra hurling rain showers. Dvipa is a hasti-yantra, as we have already noted, employed against elephants in battle; Jvara may refer to a weapon used for biological warfare. Vyāmiśra is a yantra that can be used as a substitute for all types of yantras.

कल्पवृक्षक्रियाविस्मापितदुर्जयस्यमान्धातृनाम्नःस्थपतेःप्रशस्तवास्तु-   शास्त्रार्थ-सारसामस्त्यसंहारोन्मिलितप्रयोगतन्त्र......वास्तुविस्तार   कुशलषण्णवति-प्रासादविधिविशारदो यानासनशयनादि विकल्पनापटुः स्थिर-चर-धार-द्विप-ज्वर-व्यामिश्रसंज्ञानां षडविधानां यन्त्राणां अद्वितीयप्रयोक्ता षड्त्रिंशदाचार्यगुणैः अलंकृतः ललितालयनामा समस्तसूत्रग्रहीवर्धकी तक्षकपक्षप्रतीक्ष्यः क्षत्रियैश्च संस्कृतः (सत्कृतः) स्थपतिरभ्येत्य विरचितांजलिराद्रदृष्टिर्निदिष्टायां भूमावुपाविशत् ।

अवसिता एव सर्वे नित्यप्रमादशैथिल्यानां शिल्पातिशयाः यतो अद्य त्वेवं प्रयोगलेशोऽपि   विस्मयाय   लोकस्य ।   युस्माद्दृशां   तु ब्रह्मेन्द्रपराशरप्रभृतिप्रणीतशास्त्रहृदयवेदिनां कियदिवैतन्नैपुणम् ।

kalpavṛkṣakriyāvismāpitadurjayasyamāndhātṛnāmnaḥstha pateḥprasastavāstu-   śāstrārthasārasāmastyasaṁhāronmilita prayogatantra......vāstuvistāra   kuśalasaṇṇavati prāsādavidhiviśārado   yānāsanaśayanādi   vikalpanāpaṭuḥ sthira-chara-dhāra-dvipa-jvara-vyāmiśra-saṁjñānāṁ ṣaḍavidhānāṁ   yantrāṇām   advitīyaprayoktā ṣaṭtriṁśadāchāryaguṇaiḥ alaṁkṛtaḥ lalitālayanāmā samast asūtragrahīvardhakī takṣakapakṣapratīkṣayaḥ kṣatriyaiścha saṁskṛtaḥ (satkṛtaḥ) sthapatirabhyetya virachitāṁjalir ādradṛṣṭirnidiṣṭāyāṁ bhūmāvupāviśat ।

avasitā ēva sarve nityapramādaśaithilyānāṁ śilpātiśayāḥ yato adya tvevaṁ prayogaleśo'pi vismayāya lokasya । yusmādṛśāṁ   tu   brahmendraparāśaraprabhṛtipraṇītaśāstra hṛdayavedināṁ kiyadivaitannaipuṇam ।

The machines referred to here are also mentioned by Kauṭilya (15th century B.C.) in his *Arthasāstra* and Bhoja (336-392 A.D.) in the *Samarāṅgaṇa sūtradhāra*, Chapter 31 (See Appendix).

From the following passage, it would appear that these yantras are dealt with in treatises associated with the authors Brahmā, Indra, and Parāśara and that they are no longer used due to prolonged neglect that even humble efforts in the line excited people's wonder.

In the preceding lines, we have discussed yantras dealing with weapons of war. Now we may take up those yantras used for pleasure, entertainment, household fittings, architectural engineering to reduce human labour; for sport, merriment, toys, and gadgets of-miscellaneous entertainment.

We may begin with Somadeva Sūri (3rd century A.D.), an encyclopaedic Jain writer and his long religious poem, the *Yasastilaka Champu,* written in South India in 949 A.D.

52. Cool Yantra-Dhārā-Gṛha: In the first part of the work, Somadeva describes the hero resorting to the cool *yantra-dhārā-grha* to spend the hot hours of the summer days.

53. *Kṛtrima-megha-Mandira*: The park described by Somadeva was fitted with mechanical fountains appropriately called by the commentator as K,trima-megha-Mandira, the artificial cloud-pavilion. It is erected in a dense garden in an area provided with many canals.

54. *Salila-tulikā*: In the park described by Somadeva, there is a stream for water sports, in the midst of which is a sandbank raised like a pavilion, provided with a water bed called Salila-tulikā.

55. *Jala-dhārā Yantra (artificial waterfall)*: Nearby Salila-tulikā is numerous vessels containing fragrant water at one end called a *yantra-jala-dārā*, a contrivance-producing artificial waterfall; the water is taken through and thrown out of the mouths of figures of elephants, tigers, lions, snakes, and others.

56. Artificial figures: The park also contains several artificial works that are figures of celestial trees, Kalpavrikṣas, with celestial damsels seated on them along with their lovers and figures of cloud-damsels- (payodhara-purandhri or Meghaputtalikā) giving shower-baths from their, bosoms, figures of monkeys spouting water, statuettes of water damsels (jaladevatas); there are wind-damsels (pavana-kanyakās), wafting breezes with fly-whisks.

57. Lady shaped Robots: The park also has lady shaped robots scattering cool sandal water. Somadeva Suri mentions such a robot that if her hands were touched, she would emit sprays through her nails; if her face, through the eyes and so on, a description which, when seen from the perspective of Bhoja's treatise, clearly proves to be a factual technology and not mere imaginative fiction.

58. Mechanical fountains: The mechanical fountains were constructed as a necessary adjunct to all palaces are seen even in the casual descriptions in the dramas, the *Mālavikāgnimitra* (2.12) and the *Nāgānanda* (3.7), for example, describing *jalayantras*.

बिन्दुक्षेपान पिपासुः परिसरति शिखी भ्रान्तिमद्धारियन्त्रम् ॥
यन्त्रोन्मुक्तश्च वेगाच्चलति विटपिनां पूरयन्त्रालवालान् ॥

bindukṣepāna pipāsuḥ parisarati śikhī
bhrāntimadvāriyantram ||
yantronmuktaścha vegāchchalati viṭapināṁ
pūrayannālavālān ||

59. Yantra putrikā for playing Fan: The breeze-lady in the *yantrādhikāra-gṛha* in the park has her companion within the bed-chamber where Somadeva Suri describes how near the bed was a yantra-putrikā plying a fan for the King's relief.

उपान्तयन्त्रपुत्रिकोत्क्षिप्यमाणव्यजनपवनापनीयमान सुरतश्रमः ।

*upāntayantraputrikotkṣipyamāṇavyajanapavanāpanīyamā
na surataśramaḥ ।*

In Bhoja's Samarāṅgaṇa Sūtradhāra, we find many other yantras of this class.

60. Dāru-yantras (Wooden yantras): We also have some information on smaller machines and instruments in the *Śloka saṅgraha* of Budhasvamin (Chapter 19). At Champā, when Naravāhanadatta was staying with Gandharvadattā, they heard the story of the origin of a local water festival. An old King of Champa had a Queen who desired to move about in waters filled with aquatic animals during her pregnancy. For her sake, the King dammed a river, widened it into a big lake, and fitted it with wooden replicas *(Dāru-yantras)* of crocodiles, fishes, and others, which moved freely in the water, and there let her sport in a vessel shaped like a vimāna.

तत्र नक्रादिसंस्थानदारुयन्त्रनियन्त्रिते ।
विमानाकारपातस्थौ तौ राजानौ विचेरतुः ॥

tatra nakrādisaṁsthānadāruyantraniyantrite ।
vimānākārapātasthau tau rājānau vicheratuḥ ||

61. Electronic yantras for entertainment: Of the two Kashmirian versions of the *Brihatkathā,* Kshemendra's (6th century A.D.) is brief, and Somadeva's is long enough for us to glean much information about yantras. These two works may be noticed together. We have, here, material bearing on three classes of electronic yantras, dolls and entertainment pieces, men and women robots, and vimanas.

In the course of the Madanamañchukā story in 4.3, Somadeva (11th century A.D.) narrates in his Kathāsaritsāgara the episode of Somaprabhā, the daughter of Maya, who takes a fancy to Madana-mañcukā and becomes her great friend. Somaprabhā and Svayamprabhā of Rāmāyaṇa and Pañchapsaras fame were two daughters of the Asura architect Maya. One morning Somaprabhā calls on Madana-mañchukā with a wonder box full of various kinds of electronic dolls. Maya himself had taught Somaprabhā how to make these yantras, electronic wooden toys. Then follows a description of four of these toy yantras: By striking at a pin, one makes a yantra jump up in the air, and it comes down with a garland; another similarly comes back with a cup of water; a third dances; and the fourth sits up and gossips. After entertaining Madana-mañchukā with these, Somaprabhā leaves the box in the former's care and departs. The next day, when she calls again, Madanamañchukā introduces her to her parents. The two then go to the royal park, where one of the toys brings forth a Buddha image and materials for worship. Hearing of this, the parents rush to see the wonder, and on the King asking her about the yantras and how they go into action, Somaprabhā gives a short account of the yantras that her father had devised; these yantras had a close similarity to those described Bhoja some 800 years ago. It also shows that these machines, electronic equipment, and vimanas were known to the experts until King Bhoja (336-392 A.D.).

Five categories of yantras and their functions

Somaprabhā says that, just as the universe comprises five elements, yantras are also based on the five elements of earth, water, fire, air, and ether.

The yantra based on earth materials (prithvi-pradhana) undertakes activities like shutting doors; a water-based yantra will be as lively as a living organism; a fire-yantra emits flames; an air yantra moves to and fro; and the element of Ākāśa serves to convey the sound generated by these yantras. Somaprabhā adds that there is a super-yantra called Cakra-yantra with miraculous powers that her father did not teach her. Then, with the parent's permission, Somaprabhā takes Madana-mañchukā in an aerial vehicle

(49) to her father's place, where her sister Svayamprabhā lives and returns to Takṣilā (present-day Taxila in Pakistan). Kṣemendra's account of this episode is brief, comprising about a dozen verses in his *Brihat-kathā-mañjari* (7.195-207).

Human Robots

Mention of yantras also figures in the *Ratnaprabhā-lambaka,* of *Brihat-kathā-mañjari* (7.9), the story of Karpurika in the city of Karpurasambhava, of whom NaravÈhanadatta has a dream. The Prince starts in search of her in the company of Gomukha. *En route,* they come to a Hemapura, where, as Naravāhanadatta is going along the bazaar street, he comes across everything about a city, shops, things, servants, men, and women, but their speechless movements and activities reveal to him the wondrous fact of their all being robots (7.10-11). He then makes his way to the palace, where he finds the only sentient being sitting on a throne like a king, and, like the soul presiding over the body and senses, he manipulates the mechanical city and is served on all sides by his robotic servants. On being asked by Naravāhanadatta, the mystery man of this machine city recounts his story: He, Rājyadhara by name, and his elder brother Prāṇadhara, were originally residents of Kāñchlpura where King Bahubala (Mahabala in Kṣemendra) was ruling. Both were śilpis (electronic engineers) and adepts in manufacturing electronic (magic) yantras, devised originally by Maya. It may be noted in passing that, according to Daṇḍin, Kañchi had some śilpis (mechanical engineers) who were adept in the making of yantras. The elder brother sought the company of courtesans and squandered his and his younger brother's property. Reaching the end of his material resources, the elder brother thought of theft and harnessed his skill in yantras for that purpose. He devised a pair of electronic swans that could move along an electronic rope operated from one end remotely. The other end of the rope was tied to the window of the King's treasury into which, night after night, these swans were sent; their beaks, which were put into action, removed the lid of jewel boxes and picked up some jewels and then the swans, were again moved back to their original place. This mysterious theft was going on, and the King ordered an all-night vigil to catch the culprits. The swans were seen doing this dexterous job and were detached from the rope, which suddenly sagged; the fastening nail became loose, and Prāṇadhara immediately understood that the theft had been caught.

Aerial Vehicle

When Prāṇadhāra understood that his theft had been caught, he asked his brother to accompany him to a far-off place to escape being caught as thieves early in the morning. Prāṇadhāra said that he had with him an aerial vehicle that could, in a single sweep, fly across 800 *yojanas,* and he got into it immediately with his family. As the seating capacity of that yantra had been reached by the crowd that entered it, Rajyādhara left in his vehicle, a *vāta-yantra vimāna,* as it is here styled, which he had made; that vimāna took him at one jerk over 200 *yojanas,* and with a second propelling over another 200 *yojanas.*

That brought the younger brother, the narrator, to the city of Hemapura. When he reached there, it was an abandoned place; he thought of peopling it with the help of his mechanical skill and created a yantra-population. The next day, he learnt from Naravāhanadatta the search that the latter was on and helped him with an aircraft that took him to the city of Karpūrasambhava. There the Prince found the heroine of his dream, Karpurikā, and married her. When he desired to return home with his new bride, his father-in-law revealed that he, too, had a visitor-śilpi in his city who could provide him with an aircraft. It happened that the visitor-śilpi was no other than Prāṇādhara, the elder brother of Kañchi. The vehicle that Prāṇadhara gave him was a veritable flying fortress, as it could lightly bear a thousand passengers, and in that, they returned home, touching Rājyadhara's city *en route.*

In Kshemendra's brief narration of this story (14. 459-508), the thieving swans are mentioned as many, the elder's aircraft is called a *Yantracakra,* and the younger's is said to possess double the speed of the elder's.

There are also two other contexts in the *Kathāsaritsāgara* where the aerial yantra figures (the story of Puṣkarākṣa and Vinayavatī and "the story of Somaprabhā and the three suitors.

Yantra- dhārā-gṛha (Fountain Pavilion)

In addition to the above, we come across two works of Bhoja (336-392 A.D.), one a poetic composition and the other a technical treatise. The former is the prose work of fiction, *Sriṅgāra-mañjari.* In it, Bhoja (336-392 A.D.) himself is the hero. Since it is not proper for a noble soul to indulge in self-glorification, Bhoja is represented by a *yantra-putrikā* (robotic figure) to discharge his functions. This work gives an elaborate description of a *yantra-dhārā-gṛha,* a fountain pavilion, with multi-mechanical functions: there are figures in constant action; there is a *jala-yantra-putrikā* (water spraying robot) that scatters a fine spray; *dhārās* which fall like slender lotus-stalks, and in a curve like a bow; electronic drum-players filling the place with their rhythms; an artificial lotus pond; toy bees which keep humming; *yantra-vrikṣa* or trees with monkey figures; a pond filled with replicas of cranes bending over and getting deceived by fishes which come near and move on; artificial tortoises diving and coming up now and then; and yantra-orchestras.

Samarāṅgaṇa-Sūtradhāra (SS.)

The *Samarāṅgaṇa-Sūtradhāra* ascribed to Bhoja (336-

392 A.D.) is, in many ways, a rare treatise in Sanskrit literature; besides the Arthaśāstra, it is the only theoretical text that has substantial information on various yantras; its value, however, is greater than that of the Arthaśāstra, as Bhoja (336-392 A.D.) goes into the details of the construction of these yantras and explains at the beginning the principles underlying yantras.

Chapter 31 of *Samrāṅgaṇa Sūtradhāra* by King Bhoja (336-392 A.D.) is devoted to the electronic device-making or Yantra-vidhāna. While defining the yantra, (mechanical or electronic device) he observes as under:

यदृच्छया प्रवृत्तानि भूतानि स्वेन वर्त्मना ।
नियम्यास्मिन् नयति यत् तद् यन्त्रमिति कीर्तितम् । 31.3

yadṛchchhayā pravṛttāni bhūtāni svena vartmanā /
niyamyāsmin nayati yat tad yantramiti kīrtitam / 31.3

Accordingly, an electronic device (yantra) is made by controlling and directing the functioning of bhūtas (earth, waters, air, and Agni, present in the form of heat, light, and electricity) in the desired way for a particular desired result.

स्वरसेन प्रवृत्तानि भूतानि स्वमनीषया ।
कृतं यस्माद् यमयति तद्वा यन्त्रमिति स्मृतम् ॥ 31.4

svarasena pravṛttāni bhūtāni svamanīṣayā /
kṛtaṃ yasmād yamayati tadvā yantramiti smṛtam // 31.4

According to him, an electronic device/yantra means intellectual control of bhūtas (earth, waters, gases, and agni present in the form of heat, light, and electricity) and makes them function according to one's needs and requirements. The word yantra is derived from the root *yaṃ* 'to control'.

The next topic Bhoja (336-392 A.D.) deals with is Bija. Bija means a constituent element. The constituent elements of a yantra are four: Earth, Water, Agni (heat, light, and electricity) and Air (gases),

Since Ākāśa (ether) provides shelter or base to all the four ones, the same also automatically becomes a constituent element.

तस्य बीजं चतुर्धा स्यात् क्षितिरापोऽनलोऽनिलः ।
आश्रयत्वेन चैतेषां विजयदप्युपयुज्यते ॥ 31.5

tasya bījaṁ chaturdhā syāt kṣitirāpo'nalo'nilaḥ /
āśrayatvena chaiteṣāṁ vijayadapyupayujyate // 31.5

Bhoja (336-392A.D.) then discusses whether *Suta* (mercury), which is an indispensable ingredient, is to be held as one of the Bijas (source) along with Earth, Water, etc. Some earlier writers had counted it separately as a Bija. Bhoja (SS. 31.6-8) says that mercury is essentially Pārthiva, i.e., the product of the Earth, though one might find it in the liquid state and also possessing a property causing motion, like the wind.

भिन्नः सूतश्चकै(यै) रुक्तस्ते च सम्यङ् न जानते ।
प्रकृत्या पार्थिवः सूतल(सत्र) यात् तत्र क्रिया भवेत् ॥6 ॥

पार्थिवत्वादयमतो न कदाचिद् विभिद्यते ।
द्रव्यत्वादग्निजत्व हि यद्यस्य परिकल्पयते ॥7 ॥

तदा विरोधो नैवास्य पावकेनोपपद्यते ।
गन्धाद वह्लेविरोधांच स्थिता पार्थिववता बलात् ॥8 ॥

bhinnaḥ sūtaśchakai(yai) ruktaste cha samyaṅ na jānate /
prakṛtyā pārthivaḥ sūtala(satra) yāt tatra kriyā bhavet //6 //

pārthivatvādayamato na kadāchid vibhidyate /ṁ
dravyatvādagnijatva hi yadyasya parikalpayate //7 //

tadā virodho naivāsya pāvakenopapadyate /

gandhāda vahnevirodhāṁcha sthitā pārthivavatā balāt ॥8॥

He (SS. 31.10-11) divides the yantras into two classes, *Svayam-vāhaka,* automatic and *Sakrit-prerya,* requiring occasional propelling; most yantras combine these two features. Another classification is the *Antarita* (concealed), i.e., the principle of its action and its mechanism are hidden from public view; the Vāhya (portable), i.e. the machine which can be carried from one place to another place; and the third, which is obscure, but may be interpreted as the distant or proximate, meaning thereby the place from which the machine acts.

स्वयंवाह्यमिहोत्कृष्टं हीनं स्यादितरत् त्रयम् ।
अन्यदन्तरितं वाह्यं वाह्यमन्यत् त्वदूरतः ॥ 10
स्वयं वाह्यमिहोत्कृष्टं हीनं स्यादितरत् त्रयम् ॥
तेषु शंसन्ति दूरस्थलक्ष्यं निकटस्थितम् ॥11॥

svayaṁvāhyamihotkṛṣṭam hīnaṁ syāditarat trayam ।
anyadantaritam vāhyaṁ vāhyamanyat tvadūrataḥ ॥ 10

svayaṁ vāhyamihotkṛṣṭaṁ hīnaṁ syāditarat trayam ॥
teṣu śaṁsanti dūrasthalakṣayaṁ nikaṭasthitam ॥11॥

Some move many persons and things, while others require many persons to move them (SS. 31.49).

एकं बहूनि चलयेद् बहुभिश्चाल्यतेऽपरम् ॥49॥

ēkaṁ bahūni chalayed bahubhiśchālyate'param ॥49॥

That machine is best whose principle of action is concealed, which achieves manifold purposes and which excites wonder (SS. 31.12).

यद्यु(दु) त्पन्नमलक्ष्यं यदेकं बहुषु साधकम् ।
तदन्यदपि शंसन्ति यस्माद् विस्मयकृत्त्रृणाम् ॥12॥

yadyu(du) tpannamalakṣayaṁ yadekaṁ bahuṣu sādhakam ।

tadanyadapi śaṁsanti yasmād vismayakṛnnṛṇām ||12||

After a lavish encomium on the comfort and advantages to be enjoyed through the machines, Bhoja (SS, 31.21-25, 42) proceeds to describe the constituent elements called Bijas of each variety, the *Pārthiva, Taijasa,* etc.; while all the elements may be used for a single yantra, it is to be named after the dominant constituent).

वित्तैक्यादस्य निष्पत्तिर्मोक्षश्चास्मान्न दुर्लभः ।
पार्थिवं पार्थवैर्बीजैः पार्थवं जलजन्मभिः ||21||

vittaikyādasya niṣpattirmokṣaśchāsmānna durlabhaḥ |
parthivaṁ parthavairbījaiḥ parthavaṁ jalajanmabhiḥ ||21||

तदेव तेजोजनितैस्तदेव मरुदुद्द्रवैः ।
आप्यमाप्यैस्तथा बीजैरानलरानिलैरपि ||22||

tadeva tejojanitaistadeva marududbhavaiḥ |
āpyamāpyaistathā bījairānalarānilairapi ||22||

वह्निजातेऽपि बीजं स्यात् सूतः सोऽपि च वान(नि)
पार्थिवानां भवेद् बीजमाप्यानामपि वारणे(रुणम्)
इति बीजानि सर्वेषां कीर्तितान्यखिलान्यपि ।
कुडयंकरणसूत्राणि भारगोलकपीडनम् ||25||

vahnijāte'pi bījaṁ syāt sūtaḥ so'pi cha vāna(ni)
parthivānāṁ bhaved bījamāpyānāmapi vāraṇe(ruṇam)
iti bījāni sarveṣāṁ kīrtitānyakhilānyapi |
kuḍayaṁkaraṇasūtrāṇi bhāragolakapīḍanam ||25||

भूतमेकमिहोद्रिक्तमन्यद्धीनं तोऽधिकम् ।
अन्यद्धीनतरं चान्यदेवंप्रायैर्विकल्पितैः ||42||

bhūtamekamihodriktamanyaddhīnaṁ to'dhikam |
anyaddhīnataraṁ chānyadevaṁprāyairvikalpitaiḥ ||42||

They are mentioned the materials: metals-tin, iron, copper and silver-and wood, hide and textiles; the parts and the principles: the wheels and the rotation; the suspenders and the hangings; the rods, the shafts and the caps; the tools; and the work: measuring, cutting, etc. all these are also to be included under the Bijas of a Pārthiva-yantra (SS. 31.25-27).

इति बीजानि सर्वेषां कीर्तितान्यखिलान्यपि ।
कुडयंकरणसूत्राणि भारगोलकपीडनम् ॥25॥

iti bījāni sarveṣāṁ kīrtitānyakhilānyapi /
kuḍayaṁkaraṇasūtrāṇi bhāragolakapīḍanam ॥25॥

लम्बनं लम्ब्कारे च चक्राणि विविधान्यपि ।
अयस्ताम्रं च तारं च त्रपु संवित्रमर्दने ॥26॥

lambanaṁ lambkāre cha chakrāṇi vividhānyapi /
ayastāmraṁ cha tāraṁ cha trapu saṁvitpramardane ॥26॥

काष्ठं च चर्म वस्त्रं च स्वबीजेषु प्रयुज्यते ।
उर्दकः कर्तरो यष्टिश्चक्रं भ्रमरकस्था ॥27॥

kāṣṭhaṁ cha charma vastraṁ cha svabījeṣu prayujyate /
urdakaḥ kartaro yaṣṭiśchakraṁ bhramarakasthā ॥27॥

The application of five-bijas on Pārthiva-machines (dominated by earth elements) comprises heating and boiling; of water, mixing and dissolving, pouring off and filling with water, and providing a belt of water.

Height, size, closeness, and motion towards a higher plane are spatial features in Pārthiva-yantras. The element of air is to be applied through bellows, fans, flaps, etc. (SS, 31.28-32).

श्रृङ्गावली च नाराचः स्वबीजान्यौर्वरे विदुः ।
ताप उत्तेजनं स्तोभः क्षेभश्च जलसङ्गजः ॥28॥

śṛṅgāvalī cha nārāchaḥ svabījānyaurvare viduḥ /
tāpa uttejanaṁ stobhaḥ kṣebhaścha jalasaṅgajaḥ ॥28॥

एवमाद्यग्निबीजानि पार्थिवस्य प्रचक्षते ।

धारा च जलभारश्च पयसो भ्रमणं तथा ॥29॥

ēvamādyagnibījāni pārthivasya prachakṣate |
dhārā cha jalabhāraścha payaso bhramaṇaṁ tathā ॥29॥

एवमदीनि भूजस्य जलजानि प्रचक्षते।
यथोच्छायो यथाधिक्यं यथा नीरन्ध्रतापि च ॥30॥

ēvamadīni bhūjasya jalajāni prachakṣate |
yathochchhāyo yathādhikyaṁ yathā nīrandhratāpi
cha ॥30॥

अत्यन्तमूर्ध्वगामित्वं स्वबीजान्ययसस्तथा।
मरुत् स्वभावजो गाढै ग्राहकैश्च प्रतीप्सितः ॥31॥

atyantamūrdhvagāmitvam svabījānyayasastathā |
marut svabhāvajo gāḍhai rgrāhakaiścha pratīpsitaḥ ॥31॥

दृत्याद्यैर्वीजनाद्यैश्च गजकर्णादिदिभिः कृतः।
(छ) चाणितो गालितश्छायं बीजं भवति भूभवे ॥32

dṛtyādyairvījanādyaiścha gajakarṇādibhiḥ kṛtaḥ |
(chha) chāṇito gālitaśchāyaṁ bījaṁ bhavati bhūbhave ॥32

Similarly, in machines, mainly Jala-yantras, timber, hide, and metal forms, the Pārthiva substances are used. (SS. 31.33-41).

काष्टंभृ(कृ) त्तिश्च लोहं च जलजे पार्थिवं भवेत्।
अन्यदम्भस्तदप्यस्तु तिर्यगूर्ध्वमधस्तथा ॥33॥

kāṣṭambhṛ(kṛ) ttiścha lohaṁ cha jalaje pārthivaṁ bhavet |
anyadambhastadapyastu tiryagūrdhvamadhastathā ॥33॥

बीजं स्वकीयं भवति यन्त्रेषु जलजन्मसु।
तापाद्यं पूर्वकथितं वह्निजं जलजे भवेत् ॥34॥

bījaṁ svakīyaṁ bhavati yantreṣu jalajanmasu |
tāpādyaṁ pūrvakathitaṁ vahnijaṁ jalaje bhavet ॥34॥

सङ्गृहीतश्च दत्तश्च पूरितः प्रतिनोदितः।
मरुद् बीजत्वमायाति यन्त्रेषु जलजन्मसु ॥35॥

saṅgṛhītaścha dattaścha pūritaḥ pratinoditaḥ |
marud bījatvamāyāti yantreṣu jalajanmasu ||35 ||

वह्निजातेषु मृत्ताम्रलोहरुक्मादि तद्ग्रहे ।
पार्थिवं कथयन्तीह बीजं बीजविचक्षणाः ||36 ||

vahnijāteṣu mṛttāmraloharukmādi tadgrahe |
pārthivaṁ kathayantīha bījaṁ bījavichakṣaṇāḥ ||36 ||

वह्रेर्वह्रर्भिवेद् बीजमाप आपस्तथा भवेत् ।
आद्यैर्दृत्यादिभिः प्रोक्तैर्मरुद् गच्छति बीजताम् ||37 ||

vahnervahnarbhived bījamāpa āpastathā bhavet |
ādyairdṛtyādibhiḥ proktairmarud gachchhati bījatām ||37 ||

प्रत्येकं च जनकं प्रेरकं ग्राहकं तथा ।
सङ्ग्राहकं च भूजातं बीजं स्यादनिलोद्भवैः ||38 ||

pratyekaṁ cha janakaṁ prerakaṁ grāhakaṁ tathā |
saṅgrāhakaṁ cha bhūjātaṁ bījaṁ syādanilodbhavaiḥ ||38 ||

प्रेरणं चाभिघातश्च विवर्तो भ्रमणं तथा ।
जलजं मारुतोत्थेषु बीजं स्यादिति सम्तम् ||39 ||

preraṇaṁ chābhighātaścha vivarto bhramaṇaṁ tathā |
jalajaṁ mārutotthesu bījaṁ syāditi samtam ||39 ||

सङ्गृहीतस्य तापाद्यैर्यानि पावकजन्मनि ।
प्रकीर्तितानि तान्येव भवन्ति पवनोद्भवैः ||40 ||

saṅgṛhītasya tāpādyairyāni pāvakajanmani |
prakīrtitāni tānyeva bhavanti pavanodbhavaiḥ ||40 ||

प्रेरितः सङ्गृहीतश्च जनितश्च समीरणः ।
आत्मनो बीजतां गच्छत्येवमन्यत् प्रकल्पयेत् ||41 ||

preritaḥ saṅgṛhītaścha janitaścha samīraṇaḥ |
ātmano bījatāṁ gachchhatyevamanyat prakalpayet ||41 ||

However, machines have to take some shape and possess a body; the earth is a crucial constituent (SS.31.43-44).

नाना भेदा भवन्त्येषां कस्तान् कात्स्र्येन वक्ष्यति ।
निष्क्रिया भू: क्रिया त्वंशे शेषेषु सहजा त्रिषु ॥43॥

nānā bhedā bhavantyeṣāṁ kastān kārtsnyena vakṣayati /
niṣkriyā bhūḥ kriyā tvaṁśe śeṣeṣu sahajā triṣu //43 //

अत: प्रायेण सा जन्या क्षितावेव प्रयत्नत: ।
साध्यस्य रूपशवत: सन्निवेशो यतो भवेत् ॥44॥

ataḥ prāyeṇa sā janyā kṣitāveva prayatnataḥ /
sādhyasya rūpaśavataḥ sanniveśo yato bhavet //44 //

The merits of a good machine, *yantra-gunas,* are as follows:

1. Proper, proportionate utilization of the elements constituting it.

2. Well-knit construction

3. Fineness of appearance

4. Inscrutability

5. Functional efficiency

6. Lightness

7. Freedom from noise where it is not part of the scheme

8. A loud noise when noise is intended as an end

9. Freedom from looseness.

10. Freedom from stiffness

11. Smooth and unhampered motion

12. Production of the intended effects (in cases where the ware is of the category of curios).

13. Securing the rhythmic quality in motion (particularly in entertainment wares)

14. Going into action when required

15. Resumption of the still state when not required (chiefly in cases of the pieces for pastime)

16. Freedom from an uncouth appearance.

17. Verisimilitude (in the case of bodies intended to represent birds, animals, etc.

18. Firmness

19. Softness

20. Durability (SS. 31. 45-49)..

यन्त्राणामाकृतिस्तेन निर्णेंतु नैव शक्यते ।
यथावद्वीजसंयोगः सौश्लिष्ट्यं श्लक्ष्णतापि च ॥45॥

yantrāṇāmākṛtistena nirṇemtu naiva śakyate /
yathāvadbījasamyogaḥ sauśliṣṭayam ślakṣaṇatāpi cha ॥45॥

अलक्षता निर्वहणं लघुत्वं शब्दहीनता ।
शब्दे साध्ये तदाधिक्यमशैथिल्यमगाढता ॥46॥

alakṣatā nirvahaṇam laghutvam śabdahīnatā /
śabde sādhye tadādhikyamaśaithilyamagāḍhatā ॥46॥

वहनीषु समस्तासु सौश्लिष्ट्यं चास्खलद्गति ।
यथाभीष्टार्थकारित्वं लयतालानुगामिता ॥47॥

vahanīṣu samastāsu sauśliṣṭayam chāskhladgati /
yathābhīṣṭārthakāritvam layatālānugāmitā ॥47॥

इष्टकालेऽर्थदर्शित्वं पुनः सम्यक्त्वसंवृतिः ॥
अनुलबणत्वं ताद्रूप्यं दाढर्ये मसृणता तथा ॥48॥

iṣṭakāle'rthadarśitvam punaḥ samyaktvasamvṛtiḥ ॥
anulabaṇatvam tādrūpyam dāḍharye masṛṇatā tathā ॥48॥

चिरकालसहत्वं च यन्त्रस्यैते गुणाः स्मृताः ।
एकं बहूनि चलयेद् बहुभिश्चाल्यतेऽपरम् ॥49॥

chirakālasahatvam cha yantrasyaite guṇāḥ smṛtāḥ /

ēkaṁ bahūni chalayed bahubhiśchālyate'param ॥49॥

Functions of yantras: According to the *Karman* or function, these machines are characterized not only by movement peculiar to each but also by the particular times when they are to operate. The speciality of some is sound, of some height, form or touch, and so on. Movement is across, upward, downward, backwards, forward, on either side, speeding and crawling. Another factor is the time taken for the movement. In sound, the factors are the varied quality of pleasing or the capacity to terrify. In electronic devices for pure entertainment, music, dance, drama, and imitation of different things and beings are the main factors, each, like music and dance, having its sub-varieties. In motion, going up and coming down. For the reproduction of whole-themes in-machinery, Bhoja (336-392 A.D.) gives the instance of the fight between the Devas and the Asuras, the churning of the ocean, Nṛsinha killing Hiraṇyakaśipu, races, elephant-fights, a mock-army, etc. (SS. 31.50-62).

सुश्लिष्टत्वमलक्षत्वं यन्त्राणां परमो गुणः ।
अथ कर्माणि यन्त्राणां विचित्राणि यथाविधि ॥50॥

suśliṣṭatvamalakṣatvaṁ yantrāṇāṁ paramo guṇaḥ ।
atha karmāṇi yantrāṇām vichitrāṇi yathāvidhi ॥50॥

न विस्तरान्न सङ्क्षेपात् साम्प्रतं संप्रचक्ष्महे ।
कस्यचित् सा क्रिया साध्या कालः कस्यापि कस्यचित् ॥51॥

na vistarānna saṅkṣepāt sāmpratam saṁprachakṣamahe ।
kasyachit sā kriyā sādhyā kālaḥ kasyāpi kasyachit ॥51॥

शब्दः कस्यापि चोच्छ्रायो रूपस्पर्शौ च कस्यचित् ।
क्रियास्तु कार्यस्य वशादनन्ताः परिकीर्तिताः ॥52॥

śabdaḥ kasyāpi chochchhrāyo rūpasparśau cha kasyachit ।
kriyāstu kāryasya vaśādanantāḥ parikīrtitāḥ ॥52॥

तीर्यंगूध्वमधः पृष्ठे पुरतः पार्श्वयोरपि ।
गमनं सरणं पात इति भेदाः क्रियोद्भवाः ॥53॥

tīryaṁgūdhvamadhaḥ pṛṣṭhe purataḥ pārśvayorapi ।
gamanaṁ saraṇaṁ pāta iti bhedāḥ kriyodbhavāḥ ॥53॥

कालो मुहूर्तकाष्ठाद्यैर्भिन्नो भेदैरनेकधा ।
शब्दो विचित्रः सुखदो रतिकृद् भीषणस्तथा ॥54॥

kālo muhūrtakāṣṭhādyairbhinno bhedairanekadhā ।
śabdo vichitraḥ sukhado ratikṛd bhīṣaṇastathā ॥54॥

उच्छ्रायस्तु जलस्य स्यात् क्वचिद् भूजेऽपि शस्यते ।
गीतं नृत्यं च वाद्यं च पटहो वंश एव च ॥55॥

uchchhrāyastu jalasya syāt kvachid bhūje'pi śasyate ।
gītaṁ nṛtyaṁ cha vādyaṁ cha paṭaho vaṁśa ēva cha ॥55॥

वीणा च कांस्यतालश्च तृमिला करटापि च ।
यत्किंचिदन्यदप्यत्र वादित्रादि विभाव्यते ॥56॥

vīṇā cha kāṁsyatālaścha tṛmilā karaṭāpi cha ।
yatkiṁchadanyadapyatra vāditrādi vibhāvyate ॥56॥

समस्तमपि तद् यन्त्राज्जायते कल्पनावशात् ।
नृत्ये तु नाटकं चोक्षस्ताण्डवं लास्यमेव च ॥57॥

samastamapi tad yantrājjāyate kalpanāvaśāt ।
nṛtye tu nāṭakaṁ chokṣastāṇḍavaṁ lāsyameva cha ॥57॥

राजमार्गश्च देशी च यन्त्रात् सर्वं प्रसिध्यति ।
तथा जात्यनुगाश्चेष्टा विरुद्धा यास्तु जातितः ॥58॥

rājamārgaścha deśī cha yantrāt sarvaṁ prasidhyati ।
tathā jātyanugāścheṣṭā viruddhā yāstu jātitaḥ ॥58॥

ताः सर्वा पि सिध्यन्ति सम्यग्यन्त्रस्य साधनात् ।
भूचराणां गतिर्व्योम्नि भूमौ व्योमचरागमः ॥59॥

tāḥ sarvā pi sidhyanti samyagyantrasya sādhanāt ।
bhūcharāṇāṁ gatirvyomni bhūmau
vyomacharāgamaḥ ॥59॥

चेष्टितान्यपि मर्त्यानां तथा भूमिस्पृशामिव ।
जायन्ते यन्त्रनिर्माणाद् विविधानीप्सितानि च ॥60 ॥

chesṭitānyapi martyānāṁ tathā bhūmispṛśāmiva /
jāyante yantranirmāṇād vividhānīpsitāni cha ॥60 ॥

यथासुरा जिता दैवैर्यथा निर्मथितोऽम्बुधिः ।
हिरण्यकशिपुर्दैत्यो नृसिंहेन हतो यथा ॥61 ।

yathāāsurā jitā daivairyathā nirmathito'mbudhiḥ /
hiraṇyakaśipurdaityo nṛsiṁhena hato yathā ॥61 /

धावनं हस्तियुद्धं च गजानामगडोऽपि च ।
नानाप्रकार(रा) या चेष्टा नानाधारागृहाणि च ॥62 ॥

dhāvanaṁ hastiyuddhaṁ cha gajānāmagaḍo'pi cha /
nānāprakāra(rā) yā cheṣṭā nānādhārāgṛhāṇi cha ॥62 ॥

Among fittings for utility, beauty, and sport, various types of shower fountains (*dhārā-gṛhas*), swings, pleasure chambers, mechanical carriers and servants, balls, and magic are mentioned (SS.31. 63-64).

दोलाकेल्यो विचित्रश्च तथा रतिगृहाणि च ।
चित्रा सेन(ना) च कुटयश्च स्वयंवाहकसेवकाः ॥63 ॥

dolākelyo vichitraścha tathā ratigṛhāṇi cha /
chitrā sena(nā) cha kuṭayaścha svayaṁvāhakasevakāḥ ॥63 ॥

सभाश्च विविधाकाराः सत्या मायाः प्रकल्पिताः ।
एवंप्रायाणि चान्यानि यन्त्रात् सिध्यन्ति कल्पनात् ॥64 ॥

sabhāścha vividhākārāḥ satyā māyāḥ prakalpitāḥ /
ēvaṁprāyāṇi chānyāni yantrāt sidhyanti kalpanāt ॥64 ॥

Bhoja (336-392 A.D.) also describes some of the things that can be accomplished through yantras; mention of some of them is made below:

1. **Moving multi-storey house:** Five storeys could be arranged, and the bed placed on the ground floor was

made to go up to each higher floor at the end of each watch of the night. (SS. 31.65)

विधाय भूमिकाः पंच शय्या ल्वा दिभुवि स्थिता ।
प्रतिप्रहरमन्यासु सर्पन्ती याति पंचमीम् ॥65 ॥

vidhāya bhūmikāḥ paṁcha śayyā tvā dibhuvi sthitā I
pratipraharamanyāsu sarpantī yāti paṁchamīm ॥65 ॥

2. **Moving couch:** Another entertaining device is the couch called- *Kshirābdhi-śayana,* in which the serpent-like bed goes up and down by the soft action of air, like that of the serpent's breathing (SS. 31. 68-69).

अवस्तुतोऽपि वस्तुत्वं वस्तुतोऽपि तथान्यथा ।
निःश्वासेन वियद् याति श्वासेनायाति मेदिनीम् ॥68 ॥

avastuto'pi vastutvaṁ vastuto'pi tathānyathā I
niḥśvāsena viyad yāti śvāsenāyāti medinīm ॥68 ॥

क्षीरोदमध्यगा शय्या प्रतीष्टाधः फणाभृता ।
गोलश्च सूति(च)विहितः सूर्यादीनां प्रदक्षिणम् ॥69 ॥

kṣīrodamadhyagā śayyā pratiṣṭādhaḥ phaṇābhṛtā I
golaścha sūti(cha)vihitaḥ sūryādīnāṁ pradakṣiāṇam ॥69 ॥

Hereunder we give some examples of miracles that could be worked through yantras:

Production of fire in water and vice versa: Bhoja (SS. 31.67) mentions the production of fire amid water and *vice versa.*

वह्नेश्च दर्शनं तोये वह्निमध्याज्जलोद्गतिः ॥67 ॥

vahneścha darśanaṁ toye vahnimadhyājjalodgatiḥ ॥67 ॥

Appearance and disappearance of things absent and present: Bhoja also mentions the effect of the complete disappearance of a thing present before one and the projection before one of the views of a thing not present before him (SS. 31.68). How these were done is not stated.

अवस्तुतोऽपि वस्तुलं वस्तुतोऽपि तथान्यथा ।
निःश्वासेन वियद् याति श्वासेनायाति मेदिनीम् ॥68॥

avastuto'pi vastutvaṁ vastuto'pi tathānyathā /
niḥsvāsena viyad yāti śvāsenāyāti medinīm //68 //

Chronometer: In the *Samrāṅgaṇa Sūtradhāra* (31.66-67), a kind of chronometer is described; there is a circular device in which, in a broad open vessel, there are thirty, probably ivory figures, or tooth-like pieces lying flat all along the circumference; the whole thing is revolving; in the centre is the figure of a lady, who wakes up one figure or piece for every *Nāḍikā*.

एवं प्रायाणि चित्राणि सम्यक् सिध्यन्ति यन्त्रतः ।
क्रमेण त्रिशतावर्तं स्थाले दन्ता भ्रमन्त्यसौ ॥66॥

ēvaṁ prāyāṇi chitrāṇi samyak sidhyanti yantrataḥ /
krameṇa triśatāvartaṁ sthāle dantā bhramantyasau //66 //

तन्मध्ये पुत्रिका क्लृप्ता प्रति नाडिं प्रबोधयेत् ॥ 67

tanmadhye putrikā klṛptā prati nāḍiṁ prabodhayet // 67

Another chronometer-like object is described in verses 31.70-71. There is a rider on a chariot, an elephant, or any other animal; for a fixed time, say, a *Nāḍikā,* the rider on his mount goes round and at the end of the *Nāḍikā,* the chronometer strikes.

गजादिरूपे रथिकरूपतां गमितः पुमान् ॥70॥
भ्रान्त्वा नाडिकया तस्याः पर्यन्ते हन्ति भो(यो) जनम् ।

gajādirūpe rathikarūpatāṁ gamitaḥ pumān //70 //
bhrāntvā nāḍikayā tasyāḥ paryante hanti bho(yo) janam /

दीपिकापुत्रिका क्लृप्ता क्षीणं क्षीणं प्रयच्छति ॥71॥

dīpikāputrikā klṛptā kṣīṇaṁ kṣīṇaṁ prayachchhati ||71 ||

Planetorium: An astronomical model of planetarium called *Gola* is described in the *Samrāṅgaṇa Sūtradhāra* (31. 69-70), in which there are needles and the day and night movement of planets is shown.

क्षीरोदमध्यगा शय्या प्रतीष्टाधः फणाभृता ।
गोलश्च सूति(च)विहितः सूर्यादीनां प्रदक्षिणम् ||69 ||
परिभ्राम्यत्यहोरात्रं ग्रहाणां दर्शयन् गतिम् ।

kṣīrodamadhyagā śayyā pratiṣṭādhaḥ phaṇābhṛtā |
golaścha sūti(cha)vihitaḥ sūryādīnāṁ pradakṣiāṇam ||69 ||
paribhrāmyatyahorātraṁ grahāṇāṁ darśayan gatim |

Electronic Oil lamp: Among electronic devices given in the *Samrāṅgaṇa Sūtradhāra* (31.71-72) is an electronic lamp into which, at set intervals, a mechanical figure goes on pouring oil. An additional feature of this is that the figure keeps on circumambulating to a definite musical rhythm.

दीपिकापुत्रिका क्लृप्ता क्षीणं क्षीणं प्रयच्छति ||71 ||

dīpikāputrikā klṛptā kṣīṇaṁ kṣīṇaṁ prayachchhati ||71 ||

दीपे तैलं प्रनृत्यन्ती तालगन्त्या प्रदक्षिणम् ।
यावत् प्रदीयते वारि तावत् पिबति सन्ततम् ||72 ||

dīpe tailaṁ pranṛtyantī tālagantyā pradakṣiṇam |
yāvat pradīyate vāri tāvat pibati santatam ||72 ||

Other entertainment yantras are speaking, singing and dancing birds, a dancing elephant, horse or monkey, water going up and descending, a mock fight, and others developed by the manipulation of air (SS. 31.73-78).

यन्त्रेण कल्पितो हस्ती न तद् गच्छत् प्रतीयते ।
शुकाद्याः पक्षिणः क्लृप्तास्तालस्यानुगमान्मुहुः ||73 ||

yantreṇa kalpito hastī na tad gachchhat pratīyate |
śukādyāḥ pakṣiṇaḥ klṛptāstālasyānugamānmuhuḥ ||73 ||

जनस्य विस्यकृतो नृत्यन्ति च पठन्ति च।
पुत्रिका वा गजेन्द्रो वा तुरगो कर्मटोऽपि वा ||74 ||

janasya visyakṛto nṛtyanti cha paṭhanti cha |
putrikā vā gajendro vā turago karmaṭo'pi vā ||74 ||

वलनैर्वर्तनैर्नृत्यंस्तालेन हरते मनः ||
येनैव वत्म्रना क्षेत्रं ध्रियते तेन तत्पयः ||75 ||

valanairvartanairnṛtyaṁstālena harate manaḥ ||
yenaiva vatmranā kṣetraṁ dhriyate tena tatpayaḥ ||75 ||

यात्यायाति पुनस्तद्वद् गर्तात् पुष्करिणीष्वपि।
फलके कानि(?)तिष्ठन्ति धावन्त्यनुमतानि च ||76 ||

yātyāyāti punastadvad gartāt puṣkariṇīṣvapi |
phalake kāni(?)tiṣṭhanti dhāvantyanumatāni cha ||76 ||

घातां(तं) ददति युध्यन्ते निर्यान्त्यश्रमनावृतम्।
नृत्यन्ति गायन्ति तथा वंशादीन् वादयन्ति च ||77 ||

ghātāṁ(taṁ) dadati yudhyante niryāntyaśramanāvṛtam |
nṛtyanti gāyanti tathā vaṁśādīn vādayanti cha ||77 ||

निरुद्धमुक्तस्य वशान्मरुतो यन्त्रभङ्गिभिः।
याश्चेष्टा दिव्यमानुष्यस्ता एवान्न न केवलम् ||78 ||

niruddhamuktasya vaśānmaruto yantrabhaṅgibhiḥ |
yāśchesṭā divyamānuṣyastā ēvānna na kevalam ||78 ||

Bhoja (336-392 A.D.) concludes this section with the observation that these and many more similar devices could be developed; even movements impossible in actual life are possible in yantras. Regarding revealing the actual process of making these devices, Bhoja (SS. 31.79-81) says that the constituent elements and fundamental principles have been mentioned so wise experts could easily construct these. Silence on the actual construction process is said to be for preserving the copyright of those well versed in this technology rather than because of the author's ignorance about it. This may also give a material advantage to the traditional technocrats who developed all these technologies.

दुष्करं यद्यदन्यच्च तत्तद् यन्त्रात् प्रसिध्यति ।
यन्त्राणां घटना नोक्ता गुप्तर्थं नाज्ञतावशात् ॥79॥

duṣkaram yadyadanyachcha tattad yantrāt prasidhyati |
yantrāṇām ghaṭanā noktā guptartham nājñātāvaśāt ॥79॥

तन्न हेतुरयं ज्ञेयो व्यक्ता नैते फलप्रदाः ।
कथितान्यत्र बीजानि यन्त्राणां घटना न यत् ॥80॥

tanna heturayam jñeyo vyaktā naite phalapradāḥ |
kathitānyatra bījāni yantrāṇām ghaṭanā na yat ॥80॥

तस्माद् व्यक्तीकृतेष्वेषु न स्यात् स्वार्थो न कौतुकम् ।
वस्तुतः कथितं सर्वं बीजानामिह कीर्तनात् ॥81॥

tasmād vyaktīkṛteṣveṣu na syāt svārtho na kautukam |
vastutaḥ kathitam sarvam bījānāmiha kīrtanāt ॥81॥

Another interesting statement that Bhoja (SS. 31.822;84) makes is that some of the yantras described by him are those actually seen by him (*dṛṣṭāni*). More important information is that he refers to the earlier masters.

यन्त्राणि यानि दृष्टानि कीर्तितान्यत्र तान्यपि । 31.82

yantrāṇi yāni dṛṣṭāni kīrtitānyatra tānyapi | 31.82

अग्रतश्च पुनर्ब्रूमः कथितं यत्पुरातनै । 31.84

agratāścha punarbrūmaḥ kathitam yatpurātanai | 31.84

This shows that by the time of King Bhoja (336-392 A.D.), an uninterrupted tradition of ancient Vedic technological know-how was available in this country.

Qualifications of an Engineer

After speaking again of the Bijas or source elements and the wonder and pleasure of these yantras, Bhoja (336-392 A.D.) refers to the *Sutradhāras* or chief architects who developed drawings and these yantras; their qualifications are set forth as:

1. Traditionally handed down knowledge

2. Training

3. Practice and application

4. Technical skill

5. And the availability of raw material (SS.31.87)

पारम्पर्यं कौशलं सौपदेशं
शास्त्राभ्यासो वास्तुकर्मोद्यमो धीः ।
सामग्रीयं निर्मला यस्य सोऽस्मिन्-
श्चित्राण्येवं वेत्ति यन्त्राणि कर्तुम् ॥87 ॥

pāramparyaṁ kauśalaṁ saupadeśaṁ
śāstrābhyāso vāstukarmodyamo dhīḥ /
sāmagrīyaṁ nirmalā yasya so'siṁma-
śchitrāṇyevaṁ vetti yantrāṇi kartum ॥87 ॥

Classification of Technology

Bhoja (336-392 A.D.) divides technology (*Yantra-śāstra-adhikāra*) into five types based upon movement, material, purpose, utilitarian value, and comfort provided by the technical device and observes that a technocrat who knows this will be able to produce a device of excellent quality and will earn name and fame in the world (SS. 31. 88).

चित्रैर्युक्तं ये गुणः पंचरूपं
ज्ञानन्त्येनं यन्त्रशास्त्रधिकारम् ।
ये वा कृत्स्नं योजयन्तेऽत्र
सम्यक् तेषां कीर्त्तिर्द्यां भुवं चावृणोति ॥88॥

chitrairyuktaṁ ye guṇaḥ paṁcharūpaṁ
jñānantyenaṁ yantraśāstradhikāram |
ye vā kṛtsnaṁ yojayante'tra
samyak teṣāṁ kīrttirdyāṁ bhuvaṁ chāvṛṇoti ||88 ||

Now we render hereunder a series of yantras with some details of their manufacture as mentioned in *Samrāṅgaṇa Sūtradhāra*.

1. A wooden bird: in whose hollow body is placed a copper contrivance one inch long and one-quarter inch high, of slender cylindrical shape, in two well-joined halves allowing a hole at the centre along which air passes when the bird moves, creating a pleasing sound (SS. 31. 89-90).

अङ्गुलेन मितमङ्गुलपादेनोच्छ्रितं द्विपुटकं तनुवृत्तम् ।
संविधेयमृजु मध्यगरन्ध्रं श्लिष्टसन्धि दृढताम्रमयं तत् ॥ ।89 ॥

aṅagulena mitamaṅgulapādenochchhritaṁ dviputakaṁ
tanuvṛttam |
saṁvidheyamṛju madhyagarandhraṁ śliṣṭasandhi
dṛḍhatāmramayaṁ tat || |89 ||

दारवेषु विहगेषु तदन्तः क्षिप्तमुद्गतसमीरवशेन ।
आतनोति विचलन्मृदुशब्दं शृण्वतां भवति चित्रकरं च ॥90 ॥

dāraveṣu vihageṣu tadantaḥ kṣiptamudgatasamīravaśena |
ātanoti vichalanmṛduśabdaṁ śṛṇvatāṁ bhavati
chitrakaraṁ cha ||90 ||

2. Relaxation device: The next is noted as a relaxation device in the bedroom. In the hollow of the bird mentioned above is placed a small drum-like piece in halves and with an air-passage as in the previous yantra; the interior device is to be loosely hung, and as the bird oscillates, a highly pleasing sound is created, which reduces the anger of the ladies (SS. 31. 91-92).

सुश्लिष्टखण्डद्वितयेन कृत्वा सरन्ध्रमन्तर्मुजानुकारम् ।
ग्रस्तं तथा कुण्डलयोर्युगेन मध्ये पुटं तस्य मृदु प्रदेयम् ॥91 ॥

suśliṣṭakhaṇḍadvitayena kṛtvā
sarandhramantarmujānukāram /
grastaṁ tathā kuṇḍalayoryugena madhye puṭaṁ tasya
mṛdu pradeyam //91 //

पूर्वोक्तयन्त्रे विधिनोदरेऽस्य क्षिप्तेऽथ शय्यातलसंस्थमेतत् ।
ध्वनिं ततः संचलनादनङ्गक्रीडारसोल्लासकरं करोति ॥92 ॥

pūrvoktayantre vidhinodare'sya kṣipte'tha
śayyātalasaṁsthametat /
dhvaniṁ tataḥ saṁchalanādanaṅgakrīḍārasollāsakaraṁ
karoti //92 //

3. Automatic Musical Intruments: Other bedroom accessories are various mechanical musical instruments that sound automatically on the principle of stopping and releasing air according to plan (SS. 31.93-94).

अस्मिञ् शय्यातलविनिहिते मुंचति व्यक्तरागं
चित्राञ् शब्दान् मृगशिशुदृशां यान्ति(ति) भीत्येव मानः ।
किंचैतासां दयितमभितो निर्भरप्रेमभाजां ।
प्रौढिं गच्छन्त्यधिकमधिकं मन्मथक्रीडितानि ॥93 ॥

asmiñ śayyātalavinihite mum̐chati vyaktarāgaṁ
chitrāñ śabdān mṛgaśiśudṛśāṁ yānti(ti) bhītyeva mānaḥ /
kiṁchaitāsāṁ dayitamabhito nirbharapremabhājāṁ /
prauḍhiṁ gachchhantyadhikamadhikaṁ
manmathakrīḍitāni //93 //

पटहमुरजे वेणुः शङ्खो विपंचयथ

काहला डमरुटिविले वाद्यातोद्यान्यमून्यखिलान्यपि ।
मधुरमधिकं यच्चित्रं च ध्वनि विदधात्यलं
तदिर विधिना रुद्धोन्मुक्तानिलस्य विजृम्भितम् ॥94॥

paṭahamuraje veṇuḥ śaṅkho vipaṁchyatha
kāhalā ḍamaruṭivile vādyātodyānyamūnyakhilānyapi |
madhuramadhikaṁ yachchitraṁ cha dhvani vidadhātyalaṁ
tadira vidhinā ruddhonmuktānilasya vijṛmbhitam ||94 ||

4. Aerial Vehicle: The third class of Yantra described is the aerial vehicle, to which we shall come last.

5. Robots: The fourth category comprises male and female-shaped robotic figures designed for various functions and services. Each part of these figures is made and fitted separately, with holes and pins, so that thighs, eyes, neck, hand, wrist, forearm, and fingers can act according to the need. The material used is mainly wood, but a leather cover is given to complete the impression of a human being. The movements are managed by holes, pins, and strings attached to rods controlling each limb. Looking into a mirror, playing the lute and stretching out the hand to touch, giving a pan, sprinkling water, and making obeisance (SS. 31.101-104) are the acts done by these figures.

दृग्ग्रीवातलहस्तप्रकोष्ठबाहूरुहस्तशाखादि ।
सच्छिद्रं वपुरखिलं तत्सन्धिषु खण्डशो घटयेत् ॥101 ॥

dṛggrīvātalahastaprakoṣṭhabāhūruhastaśākhādi |
sachchhidraṁ vapurakhilaṁ tatsandhiṣu khaṇḍaśo
ghaṭayet ||101 ||

श्लिष्टं कीलकविधिना दारुमयं सृष्टचर्मणा गुप्तम् ।
पुंसोऽथवा युवत्या रूपं कृत्वातिरमणीयम् ॥102 ॥

śliṣṭaṁ kīlakavidhinā dārumayaṁ sṛṣṭacharmaṇā guptam |
puṁso'thavā yuvatyā rūpaṁ kṛtvātiramaṇīyam ||102 ||

रन्ध्रगतैः प्रत्यङ्ग विधिना नाराचसङ्गतैः सूत्रैः ।
ग्रीवाचलनप्रसरणविकुंचनादीनि विदधाति ॥103॥

randhragataiḥ pratyaṅga vidhinā nārāchasaṅgataiḥ sūtraiḥ /
grīvāchalanaprasaraṇavikuṁchanādīni vidadhāti ॥103॥

करग्रहणताम्बूलप्रदानजलसेचनप्रकाणा(णामा)दि ।
आदर्शप्रतिलोकनवीणावाद्यदि च करोति ॥104॥

karagrahaṇatāmbūlapradānajalasechanaprakāṇā(ṇāmā)di /
ādarśapratilokanavīṇāvādyadi cha karoti ॥104॥

It is one such that provides the mechanical fan in the *Yasastilaka Champu.*

Similar robots are used for the palace guard; one such is made to stand at the gate with a baton, sword, iron rod, spear, or other weapon and prevent the entry of outsiders. This can quickly and quietly kill thieves who break into the palace at night (SS. 31.106-107).

पुंसो दारुजमूर्ध्वं रूपं कृत्वा निकेतनद्वारी ।
तत्करयोजितदण्डं निरुणद्धि प्रविशतां वर्त्म ॥106॥

puṁso dārujamūrdhvaṁ rūpaṁ kṛtvā niketanadvārī /
tatkarayojitadaṇḍaṁ niruṇaddhi praviśatāṁ vartma ॥106॥

खड्गहस्तमथ मुद्रहस्तं कुन्तहस्तमथवा यदि तत् स्यात् ।
तन्निहन्ति विशतो निशि चौरान् द्वारि संवृतमुखं प्रसभेन ॥107॥

khaḍgahastamatha mudgrahastaṁ kuntahastamathavā yadi tat syāt /
tannihanti viśato niśi chaurān dvāri saṁvṛtamukhaṁ prasabhena ॥107॥

6. Equipments for military: The fifth category of yantras is military equipment in forts, bows, *śataghnis* and a weapon newly mentioned by Bhoja (336-392 A.D.), the Uṣṭra-grīva (camel's neck), resembling probably the modern cranes. He also indicates that these devices were made to protect the fort *(guptyartha)* and the sports and entertainment of the kings *(Kṛḍārtha)* (SS. 31.108).

ये चापाद्या ये शतघ्न्यादयोऽस्मि-
न्नुष्ट्रग्रीवाद्याश्च दुर्गस्य गुप्त्यै ।
ये क्रीडाद्याः क्रीडनार्थं च राज्ञां
सर्वेऽपि स्युर्योगतस्ते गुणानाम् ॥108 ॥

ye chāpādyā ye śataghnyādayo'smi-
nnuṣṭragrīvādyāścha durgasya guptyai |
ye krīḍādyāḥ krīḍanārtham cha rājñām
sarve'pi syuryogataste guṇānām ||108 ||

7. Equipment of sports and entertainment: In the sixth series, the yantras are mentioned for the purposes of the kings' sports and entertainment *(Kṛḍārtha)*. They include a fountain, *vāri-yantra,* described in some detail by Somadeva and dealt with by Bhoja (336-392 A.D.) not only here but also earlier under palace architecture. Movement in the *vāri-yantra* is fourfold:

(a) **Waterfall-Machine:** A downward flow from an overhead tank for which a *Pāta-yantra,* or waterfall-machine, is to be used;

(b) ***Samanāḍikā*** is for the release of water at a higher level from tanks placed at a lower level;

(c) ***Pāta sama-ucchrāya*** is an instrument using bored columns for letting down water from a height and then taking it up through columns placed aslant; and

(d) the last, ***Ucchrāya,*** in which water from a well or in a canal on the ground is lifted with the help of a device. (S.S. 31.110-14).

निस्रगं भवति द्रोणीदेशादूर्ध्वस्थिताज्जलम् ।
यत्र तत् पातयन्त्रः स्याद् वाटिकादिप्रयोजनम् ॥110 ॥

nisnagaṁ bhavati droṇīdeśādūrdhvasthitājjalam |
yatra tat pātayantraḥ syād vāṭikādiprayojanam ॥110 ॥

उच्छ्रायसमपाताख्यं यत्रोर्ध्वा नाडिका पयः ।
जलाधारगुणान्मुंचेदधस्तात् समनाडिका(कम्) ॥111 ॥

uchchhrāyasamapātākhyaṁ yatrordhvā nāḍikā payaḥ |
jalādhāraguṇānmuṁchedadhastāt samanāḍikā(kam) ॥111 ॥

यत्र पातसमुच्छ्रायं पतित्वोच्छ्रायतो जलम् ।
तिर्यग् गत्वा पर्यात्यूर्ध्वं सच्छिद्रस्तम्भयोगतः ॥112 ॥

yatra pātasamuchchhrāyaṁ patitvochchhrāyato jalam |
tiryag gatvā paryātyūrdhvaṁ
sachchhidrastambhayogataḥ ॥112 ॥

पतित्वोच्छ्रायतस्तोयं तिर्यग्गूर्ध्वोर्ध्वमेत्यथ ।
सच्छिद्रस्तम्भयोगेन तत् स्यात् पातसमोच्छ्रयम् ॥113 ॥

patitvochchhrāyatastoyaṁ tiryagūrdhvordhvametyatha |
sachchhidrastambhayogena tat syāt
pātasamochchhrayam ॥113 ॥

वाप्यां वापि च कूपे विधानतो दीर्घिकादिका विहिता ।
यत्रोर्ध्वमम्बु गमयति तदिहोच्छ्रयसंज्ञितं कथितम् ॥114 ॥

vāpyāṁ vāpi cha kūpe vidhānato dīrghikādikā vihitā |
yatrordhvamambu gamayati tadihochchhrayasaṁjñitaṁ
kathitam ॥114 ॥

(e). **Wooden Elephant:** An artificial object which was probably common in Bhoja's time is a "wooden elephant", which Bhoja (336-392 A.D.) describes twice; earlier, he cited it as an example of various excellent effects that could be achieved through yantras (SS. 31.73).

यन्त्रेण कल्पितो हस्ती न तद् गच्छत् प्रतीयते ।
शुकाद्याः पक्षिणः क्लृप्तास्तालस्यानुगमान्मुहुः ॥73 ॥

yantreṇa kalpito hastī na tad gachchhat pratīyate /
śukādyāḥ pakṣiṇaḥ klṛptāstālasyānugamānmuhuḥ ॥73 ॥

It occurs again under yantras based on the principle of lifting water, *Ucchrāya* (SS. 31.115). This wooden-elephant drinks water placed in a vessel; any amount of it, and neither the intake nor the water is taken in, is perceivable.

दारुजमिभस्य रूपं यत् सलिलं पात्रसंस्थितं पिबति ।
तन्माहात्म्यं निगदितमेतस्योच्छ्रायतुल्यस्य ॥115 ॥

dārujamibhasya rūpaṁ yat salilaṁ pātrasaṁsthitaṁ pibati /
tanmāhātmyaṁ nigaditametasyochchhrāyatulyasya ॥115 ॥

The *Samaucchrāya* principle of water circulation on the same level is based on the underground conduit, which brings water to a tank from a distant source (SS. 31.116). Water conduits, in general, are described earlier also in Chapter 18 in connection with the city and residence.

सलिलं सुरङ्गदेशानीतं निम्नेन वर्त्मना दूरे ।
अद्भुतमम्भस्थानं तदिह समोच्छ्रायतः कुरुते ॥116 ॥

salilaṁ suraṅgadeśānītaṁ nimnena vartmanā dūre /
adbhutamambhasthānaṁ tadiha samochchhrāyataḥ
kurute ॥116 ॥

In the class of up-and-down lift of water, Bhoja (336-392 A.D.) expatiates upon the *Dhārāgṛha*, shower-bower, in the garden; its popularity has already been noted, and Bhoja has already described it in Chapter 18. Its great vogue can also be seen from its different types known by distinct names mentioned by Bhoja (SS.31.117-118):

(1) *Pravarṣaṇa,* the shower.

(2) *Pranāla,* the pipe.

(3) *Jalamagna,* the subaquatic, and

(4) The *Nandyāvarta,* in a special design.

These were manufactured only in palaces for the King's entertainment.

धारागृहमेकं स्यात् प्रवर्षणाख्यं ततो द्वितीयं च ।
प्राणालं जलमग्नं नन्द्यावर्तं तथान्यदिप ॥117 ॥

dhārāgṛhamekaṁ syāt pravarṣaṇākhyaṁ tato dvitīyaṁ cha ।
prāṇālaṁ jalamagnaṁ nandyāvartaṁ tathānyadipa ॥117 ॥

प्राकृतजनार्थमेतन्न विधेयं योग्यमेतदवनिभुजाम् ॥
मङ्गल्यानां सदनं दिव्यमिदं तुष्टिपुष्टकरम् ॥118 ॥

prākṛtajanārthametanna vidheyaṁ
yogyametadavanibhujām ॥
maṅgalyānāṁ sadanaṁ divyamidam tuṣṭipuṣṭakaram ॥118 ॥

Regarding their construction, Bhoja (SS. 31.119-120) says:

(1) They are to be in the proximity of big reservoirs;

(2) They should occupy a site with good scenic beauty;

(3) Pipes have to be prepared to double and treble the height and other requirements of the fountains; the pipes should be able to carry water, be free from pores, and smooth inside.

सलिलाशयस्य सविधे कस्याप्याश्रित्य शोभनं देशम् ।
यन्त्रोत्सेधाद् द्विगुणा त्रिगुणा वा नाडिका कार्या ॥119 ॥

salilāśayasya savidhe kasyāpyāśritya śobhanaṁ deśam ।
yantrotsedhād dviguṇā triguṇā vā nāḍikā kāryā ॥119 ॥

जलनिर्वाहसहासावन्तर्मसृणा बहिश्च नीरन्ध्रा ।
निर्व्यूढाम्भसि तस्यां शुभे मुहूर्ते गृहं कार्यम् ॥120॥

jalanirvāhasahāsāvantarmasṛṇā bahiścha nīrandhrā /
nirvyūḍhāmbhasi tasyāṁ śubhe muhūrte gṛhaṁ
kāryam ॥120॥

Naturally, architectural erections add to the excellence of the fountain park and parts of the structure are used for the different waterworks. Fine and fragrant timber, Devadāru, Sandal, and Sāl, are to be used for the woodwork, carved pillars, platforms, projections, windows, cornices, etc. main items are robotic female figures and models of birds, animals like monkeys, manifold forms with gaping mouths, semi-divine and half-animal forms, *Nāgas, Kinnaras,* etc., dancing peacocks, *Kalpavṛkṣas,* creepers and bowers, cuckoos, bees, and swans. In the centre of the flowing stream is to be fixed the main pipe, the exterior of it being made into any charming form according to one's liking. To the top of it is fitted and fastened firmly with *vajralepa,* and cement, the devices for taking up water, scattering, and throwing it in a variety of ways (SS. 31.133).

उच्छ्रययन्त्रेणेतद् भ्रान्तजलेनाथ तदभितः कृत्वा ।
चित्रानुपातयुक्तं प्रदर्शयेन्नृपतये स्थपतिः ॥133॥

uchchhrayayantreṇetad bhrāntajalenātha tadabhitaḥ kṛtvā /
chitrānupātayuktaṁ pradarśayennṛpataye sthapatiḥ ॥133॥

The pond is to be filled, for effect, with yantras of animals and aquatic beings, *e.g.,* sporting elephants, which do even minute actions like closing their eyes when another throws water on the face (SS. 31.134).

कार्याण्यस्मिन् करिणां मिथुनान्यभितोऽम्बुकेलियुक्तानि ।
अन्योन्यपुष्कारोज्झतसीकरभयपिहितनयनानि ॥134॥

kāryāṇyasmin kariṇāṁ mithunānyabhito'mbukeliyuktāni |
anyonyapuṣkāroṁjhratasīkarabhayapihitanayanāni ||134 ||

Other specimens we saw in the description in the author's *Sringaramañjari*. Female figures spraying water from eyes, nails, etc., when those parts are touched are described here (SS. 31.136-37)

स्तनयोर्युगेन सृजी जलधारे तत्र कापि कार्या स्त्री ।
आनन्दाश्रुलवानिव सलिलकणान् पक्ष्मभिः काचित् ॥136 ॥

stanayoryugena sṛjī jaladhāre tatra kāpi kāryā strī |
ānandāśrulavāniva salilakaṇān pakṣamabhiḥ kāchit ||136 ||

नाभिहृदनदिकामिव विनिर्गतां कापि बिभ्रती धाराम् ।
काप्यङ्गलीनखांशुभिरिव योषित् सिंचती कार्या ॥137 ॥

nābhihradanadikāmiva vinirgatāṁ kāpi bibhratī dhārām |
kāpyangalīnakhāṁśubhiriva yoṣit siṁchatī kāryā ||137 ||

This is also described in the earlier chapter (SS.18. 47-50).

The King's seat is right in the centre on a fine stone; he sometimes indulges in a bath, sometimes enjoys the play of water from these manifold devices, the *Jala-śilpas*, sometimes listens to music and watches dancing here, and, particularly in summer, the fountain is a necessity (SS. 31.139-141).

मध्ये तस्य विधेयं सिंहासनममलहेममणिघटितम् ।
तत्रासीदेन्नरपतिरवनिपतिः श्रीपतिर्देवः ॥139 ॥

madhye tasya vidheyaṁ
siṁhāsanamamalahemamaṇighaṭitam |
tatrāsīdennarapatiravanipatiḥ śrīpatirdevaḥ ||139 ||

स्नायात् कदाचिदस्मिन् मङ्गलगीतैर्विवर्धितानन्दः ।
वादित्रनदाट्यनिपुणैर्निषेव्यमाणः सुरेन्द्र इव ॥140 ॥

snāyāt kadāchidasmin maṅgalagītairvivardhitānandaḥ |

vāditranadātyanipuṇairniṣevyamāṇaḥ surendra iva ||140 ||

य एतस्मिन् गाढग्लपितघनघर्मव्यतिकरे
शुचौ धाराधाम्नि स्फुटसलिलधारे नरपतिः ।
सुखेनासे पश्यन् विविधजलशिल्पानि स
भवेन्न मर्त्यः किन्त्वेष क्षितिकृतनिवासः सुरपतिः ||141 ||

*ya ētasmin gāḍhaglapitaghanagharmavyatikare
śuchau dhārādhāmni sphuṭasaliladhāre narapatiḥ |
sukhenāse paśyan vividhajalaśilpāni sa
bhavenna martyaḥ kintveṣa kṣitikṛtanivāsaḥ surapatiḥ ||141 ||*

More specific descriptions of the four types of *Dhārāgṛha* are given below:

The primary speciality of the first, the *Pravarṣaṇa,* the shower, is that it pours down water. Strong figures of three, four, or seven men should be set up with curved tubes; the whole mechanism is fitted with water which is poured out in different ways by these figures (SS. 31.142-46).

जलदकुलाष्टकयुक्तं पूर्ववदन्यद् गृहं समारचयेत् ।
वर्षद्वारानिकरैः पर्वर्षणाख्यां तदाप्नोति ||142 ||

*jaladakulāṣṭakayuktaṁ pūrvavadanyad gṛhaṁ
samārachayet |
varṣadvārānikaraiḥ parvarṣaṇākhyāṁ tadāpnoti ||142 ||*

प्रतिकुलमस्मिन् कार्या दिव्यालङ्कारधारिणः पुरुषाः ।
विधिना त्रयः सुरूपाश्चत्वारः सप्त वा सुदृढाः ||143 ||

*pratikulamasmin kāryā divyālaṅkāradhāriṇaḥ puruṣāḥ |
vidhinā trayaḥ surūpāśchatvāraḥ sapta vā sudṛḍhāḥ ||143 ||*

यन्त्रेण समोच्छ्रायेण तांश्चतुर्थेन वा ततः पुरुषान् ।
कृत्वा सवक्रनालानम्भोभिः पूरयेद् विमलैः ||144 ||

*yantreṇa samochchhrāyeṇa tāṁśchaturthena vā tataḥ
puruṣān |*

kṛtvā savakranālānambhobhiḥ pūrayed vimalaiḥ ||144||

सलिलप्रवेशरन्ध्राण्यखिलानि पिधाय तत्र पुरुषाणाम् ।
अङ्गानि वारिमोक्षाण्यखिलन्यथ मोचयेत् तेषाम् ॥145॥

*salilapraveśarandhrānyakhilāni pidhāya tatra puruṣāṇām |
anagāni vārimokṣānyakhilanyatha mochayet teṣām* ||145||

सलिलं सवक्रनालं द्वारप्रतिरोधमोचनैः पुरुषाः ।
मुंचन्ति स्वेच्छममी विचित्रपातेन चित्रकरम् ॥146॥

*salilam savakranālam dvārapratirodhamochanaiḥ puruṣāḥ |
mumchanti svechchhamamī vichitrapātena
chitrakaram* ||146||

Bhoja calls this shower-house a pseudo-cloud, *anukaraṇam ekam jala-mucām* (SS. 31.148).

इदं नानाकारं कुलभवनमाद्यं रतिपतेनिर्वासि-
श्रित्राणामनुकरणमेकं जलमुचाम् ।
पयःपातैर्ग्रीष्मे रविकरपरीतापशमनं
न केषामत्थर्यं भवति नयनानन्दजननम् ॥148॥

*idam nānākāram kulabhavanamādyam ratipatenirvāsa-
śchitrāṇāmanukaraṇamekam jalamuchām |
payaḥpātairgrīṣme ravikaraparītāpaśamanam
na keṣāmattharyam bhavati nayanānandajananam* ||148||

Somadeva Suri's commentator gives it the name *kṛtrima-megha-mandira*)-a boon in summer and a feast to the eyes. Kālidāsa's reference to the yantra-dhārā-gṛha has already been mentioned, but when he says in his *Megha-sandeśa* (I. 61) *nesyanti tvām sura-yavalayo yantra-dhārā-gṛhatvam*), that the celestial damsels on the Himalayas would scratch the cloud with their bangles and convert it into a *yantra-dhārā-gṛha*, he seems to know also the name of this type called after the cloud.

The next variety called *Praṇāla* is two-storeyed with a

single pillar or four, eight or sixteen, built like a *Puṣpaka-vimāna* with decorative designs. At the centre below is a water tank with a big lotus, its pericarp fashioned as the seat of the King; around are female figures looking at the lotus; when the overhead tank is filled and closed, water is poured by the figures on the King sitting on the lotus seat.

The third, *Jalamagna,* is a chamber underwater, the idea being that of the submarine abode of Varuṇa or Nāgarāja. A square chamber is built at the bottom of an extensive and deep water reservoir, the approach to it being through a subterranean passage. A continuous flow of water above keeps the chamber wholly cool, and the reservoir is full of electronic lotuses, fish, birds, etc. When resting in this chamber alone or in a private company, the King can be seen only by select personal friends and urgent visitors of rank like other Princes or Ambassadors (SS. 31.157-166).

चतुरश्रातिगभीरा वापी कार्या मनोरमा सुदृढा ।
गर्भगतं गृहमस्याः कर्तव्यं लिप्तसन्धि ततः ॥157॥

chaturaśrātigabhīrā vāpī kāryā manoramā sudṛdhā /
garbhagataṁ gṛhamasyāḥ kartavyaṁ liptasandhi
tataḥ ॥157॥

विहितप्रवेशनिर्गति सुरङ्ग्याधो निवेशितद्वारम् ।
विदधीत चारुरूपैः प्रवर्षकैर्व्यसमुरष्टिात् ॥158॥

vihitapraveśanirgati suraṅgayādho niveśitadvāram /
vidadhīta chārurūpaiḥ pravarṣakairvyaptamuraṣṭiāt ॥158॥

चित्राध्यायोदितवर्त्मना ततोऽलङ्कृतं च चित्रेण ।
तस्य विधेयं मध्यं सलिलधिपवाससङ्काशम् ॥159॥

chitrādhyāyoditavartmanā tato'laṅkṛtaṁ cha chitreṇa /
tasya vidheyaṁ madhyaṁ saliladhipavāsasaṅkāśam ॥159॥

ऊर्ध्वविनिर्गमिताब्जैनालैस्तत्पट्टकन्दकोद्धूतैः ।
सच्छिद्रकर्णिकागतदिनकरकरनिर्मितोद्द्योतम् ॥160॥

ūrdhvavinirgamitābjainālaistatpaṭṭakandakodbhūtaiḥ ।
sachchhidrakarṇikāgatadinakarakaranirmitoddyotam ॥160

आपूरयेत् ततोऽनु च पाताम्बुभिरमलकमलपर्यन्तम् ।
विधिनामुनैव सम्यक् प्रविधाय मनोरमं भवनम् ॥161॥

āpūrayet tato'nu cha pātāmbubhiramalakamalaparyantam ।
vidhināmunaiva samyak pravidhāya manoramaṁ
bhavanam ॥161 ॥

नानारूपकयुक्त्याउ(व्यु)परचिततमङ्गतोरणद्वारम् ।
शालाभिरायताभिश्चतसृष्वपि दिक्षु कृतशोभम् ॥162॥

nānārūpakayuktyāu(vyu)parachitatamaṅgatoraṇadvāram ।
śālābhirāyatābhiśchatasṛṣvapi dikṣu kṛtaśobham ॥162 ॥

कृत्रिमशफरीमकारीपक्षिभिरपि चाम्बुसम्भवैर्युक्ताम् ।
कुर्यादम्भोजवतीं वापीमाहार्य्ययोगेन ॥163॥

kṛtrimaśapharīmakārīpakṣibhirapi
chāmbusambhavairyuktām ।
kuryādambhojavatīṁ vāpīmāhāryayogena ॥163 ॥

सामन्तमुख्यपुरुषा रारजाज्ञालब्धसंश्रयास्तत्र ।
परराष्ट्रगतदूतास्तिष्ठेयुर्नि हितमिह निभृताः ॥164॥

sāmantamukhyapuruṣā rārajājñālabdhasaṁśrayāstatra ।
pararāṣṭragatadūtāstiṣṭheyurni hitamiha nibhṛtāḥ ॥164 ॥

अथ स यथाविधि सलिलक्रीडां पूर्वोक्तमार्गरूपाणाम् ।
दृष्ट्वा मुदितः कुर्यात् पर्यङ्कारोहणं नृपतिः ॥165॥

atha sa yathāvidhi salilakrīḍāṁ pūrvoktamārgarūpāṇām ।
dṛṣṭavā muditaḥ kuryāt paryaṅkārohaṇaṁ nṛpatiḥ ॥165 ॥

तत्र स्थितस्य नृपतेः परिवारितस्य
वाराङ्गनाभिरभितो जलमग्नधाम्नि ।
पातालसद्मनि यथा भुजगेश्वरस्य

निस्सीमसम्भृतरतिर्भवति प्रमोदः ॥166॥

tatra sthitasya nṛpateḥ parivāritasya
vārāṅganābhirabhito jalamagnadhāmni I
pātālasadmani yathā bhujageśvarasya
nissīmasambhṛtaratirbhavati pramodaḥ ॥166 ॥

The last type, *Nandyāvarta,* has, in mid-tank, a prominent flower-like structure; all around the central floral design, in mid-water, are placed low walls in Svastika designs, providing a sufficient screen as well as a passage, the purpose being to permit playing in the water the game of hide-and-seek (SS. 31.167-72).

पूर्वोक्तवापिकायां मध्ये स्तम्भैश्चतुर्भिरूपरचितम्।
मुक्ताप्रवालयुक्तं पुष्पकमथ कारयेल्लटभम् ॥167॥

pūrvoktavāpikāyāṁ madhye
stambhaiśchaturbhirūparachitam I
muktāpravālayuktaṁ puṣpakamatha kārayellaṭabham ॥167 ॥

वापीं परितः पुष्पकमापूर्य सुनिर्गमाभिरथ सुदृढम्।
गभस्वस्तिकभित्तिभिरूपहितशोभं समन्ततः कुर्यात् ॥168॥

vāpīṁ paritaḥ puṣpakamāpūrya sunirgamābhiratha
sudṛḍham I
gabhasvastikabhittibhirūpahitaśobhaṁ samantataḥ
kuryāt ॥168 ॥

पूर्वोक्तवारियोगात् पूर्णामाकर्णतो विधायैताम्।
जलकेलिषु सोत्क्ण्ठो महीपतिः पुष्पकं यायात् ॥169॥

pūrvoktavāriyogāt pūrṇāmākarṇato vidhāyaitām I
jalakeliṣu sotkṇṭho mahīpatiḥ puṣpakaṁ yāyāt ॥169 ॥

कुर्वीत नर्मसचिवैर्विलासिनीभिश्च सार्धमवनिपतिः।
तद्भित्त्यन्तरवर्ती निमज्जनोन्मज्जनैः क्रीडाम् ॥170॥

kurvīta narmasachivairvilāsinībhiścha sārdhamavanipatiḥ I
tadbhittyantaravartī nimajjanonmmajjanaiḥ krīḍām ॥170 ॥

एकत्र मग्नैरपत्र दृष्टैरन्यत्र हत्वा सलिलेन नष्टैः ।
क्रीडत्यलं केलिकरः सहायैनृपः सुखं मज्जनपुष्करिण्याम् ॥171॥

*ēkatra magnairapatra dṛṣṭairanyatra hatvā salilena naṣṭaiḥ /
krīḍatyalaṁ kelikaraḥ sahāyainṛpaḥ sukhaṁ
majjanapuṣkariṇyām ||171 ||*

वापीतलस्थितमथ त्रपयावनम्रमाच्छादितस्तनभरं करपल्लेवन ।
गाढावसक्तवसनं जलरोध्मुक्तावालोकते प्राणयिनीजनमत्र धन्यः ॥172॥

*vāpītalasthitamatha
trapayāvanamramāchchhāditastanabharaṁ karapallevana /
gāḍhāvasaktavasanaṁ jalarodhmuktāvālokate
prāṇayinījanamatra dhanyaḥ ||172 ||*

The fifth central division of the *yantrādhikāra* was mentioned at the beginning as *Rathadolā*. Bhoja (336-392 A.D.) now takes it up. *Rathadolā* is a swing or a merry-go-round in which people ride in seats and enjoy the pleasure of wheeling around. Those merry-go-rounds are also seen in descriptions like the one we find in the poem *Citrabandha Rāmāyaṇa* of Venkaṭesa kavi (Tanjore Ms No. 3772, Verse 6), where the poet describes the courtezans wheeling round in the Dāru-yantra in the palace courtyard as stars going around Mount Meru. Bhoja devotes as much attention to it as to the *yantra-dhārā-gṛha*; here, too, types are known with distinct names. The varieties are called *Vasanta, Madanotsava, Vasanta-tilakā, Vibhrāmaka,* and *Tripura,* (31.174), and each subsequent type is more elaborate and complicated in its mechanism than the previous one. In the *Vasanta* type, the yantra is planted in a dugout eight cubits square and four cubits deep; metal and woodwork are mentioned at the base of the yantra, where the rotation mechanism is fitted to a platform. A storey is to be raised on twelve posts; on the whole, five machines are to be employed for the rotation, the gear wheel acting upon another gear wheel, and the whole moving the storey, designed like a lotus and accommodating the whirling riders (SS. 31.175-87).

खिनेच्चतुरः स्तम्भान् समैकसूत्रोपगान् ऋजून् सुदृढान् ।
सदृशान्तरान् धरित्रीवशतः सुश्लिक्ष्ण(ष्ट)पीठगतान् ॥175 ॥

khinechchaturaḥ stambhān samaikasūtropagān rjūn sudrḍhān /
sadrśāntarān dharitrīvaśataḥ suślikṣaṇa(ṣṭa)pīṭhagatān ॥175 ॥

प्रासादस्योक्तदिशि प्रविदध्याद् विरचिताष्टकरदैर्ध्यम् ।
भूमिगृहं रमणीयं तदर्धतो विहितगाम्भीर्यम् ॥176 ॥

prāsādasyoktadiśi pravidadhyād virachitāṣṭakaradairdhyam /
bhūmigrham ramaṇīyam tadardhato vihitagāmbhīryam ॥176 ॥

तद्गर्भतले स्तम्भो लोहमयाधारसंस्थितः कार्यः ।
भ्रमसहितः पीठयुतो ग्रस्तश्चच्छादकतुलाभिः ॥177 ॥

tadgarbhatale stambho lohamayādhārasaṁsthitaḥ kāryaḥ /
bhramasahitaḥ pīṭhayuto grastaśchachchhādakatulābhiḥ ॥177 ॥

संस्थाप्योपरि पीठस्य कुम्भिकामतिदृढां विभक्तां च ।
धनुरुच्छ्रितस्ततोऽमूमष्टभिरावेष्टयेद् भद्रैः ॥178 ॥

saṁsthāpyopari pīṭhasya kumbhikāmatidrḍhāṁ vibhaktāṁ cha /
dhanuruchchhritastato'mūmaṣṭabhirāveṣṭayed bhadraiḥ ॥178 ॥

स्वेच्छमथ भूमिकोच्छ्रयमस्योर्ध्वे कल्पयेन्नितान्तमृजुम् ।
निदधीत वेष्टनोर्ध्वे पट्टयुतं स्तम्भशीर्ष च ॥179 ॥

svechchhamatha bhūmikochchhrayamasyordhve kalpayennitāntamrjum /
nidadhīta veṣṭanordhve paṭṭayutaṁ stambhaśīrṣaṁ

cha ||179 ||

हीरग्रह(ण?)पर्यन्तं मदला गजशीर्षिका विधातव्या ।
सुदृढा प्रयत्नरचिता मनोभिरामा यथाशोभम् ||180 ||

hīragraha(ṇa?)paryantaṁ madalā gajaśīrṣikā vidhātavyā |
sudṛḍhā prayatnarachitā manobhirāmā yathāśobham ||180 ||

पट्टस्योपरि कार्या चतुष्किका क्षेत्रमानतोऽभीष्टात् ।
तस्यामुपरि विधेयस्तलबन्धो दृढतरन्यासः ||181 ||

paṭṭasyopari kāryā chatuṣkikā kṣetramānato'bhīṣṭāt |
tasyāmupari vidheyastalabandho dṛḍhataranyāsaḥ ||181 ||

स्तम्भैर्द्वादशभिरथ क्षेत्रे युक्त्या समुच्छ्रितैर्भव्यैः ।
रूपवतीकोणस्थितिराधिका भूः प्रथमिका कार्या ||182 ||

stambhairdvādaśabhiratha kṣetre yuktyā
samuchchhritairbhavyaiḥ |
rūpavatīkoṇasthitirādhikā bhūḥ prathamikā kāryā ||182 ||

मध्ये भ्रमश्च तस्या गर्भस्तम्भप्रतिष्ठितः कार्यः ।
क्षेत्रप्रमाणवशतस्तां पश्चाच्छादयेत् पट्टैः ||183 ||

madhye bhramaścha tasyā garbhastambhapratiṣṭhitaḥ
kāryaḥ |
kṣetrapramāṇavaśatastāṁ paśchāchchhādayet paṭṭaiḥ ||183 ||

रथिकाशिखाग्रकेषु च फलकाम(व) रणस्य तद्वदुपरिष्टात् ।
भ्रमचक्राणि न्यस्येन्मध्ये स्तम्भे च पंचैव ||184 ||

rathikāśikhāgrakeṣu cha phalakāma(va) raṇasya
tadvadupariṣṭāt |
bhramachakrāṇi nyasyenmadhye stambhe cha
paṁchaiva ||184 ||

अत उपरि यथाशोभं हि भूमिका पुष्पकाकृतिः कार्या ।
मध्यस्तम्भाधारा कृतकलशविभूष्णा शिरसि ||185 ||

ata upari yathāśobhaṁ hi bhūmikā puṣpakākṛtiḥ kāryā |
madhyastambhādhārā kṛtakalaśavibhūṣṇā śirasi ||185 ||

स्तम्भेऽव(ध)स्ताद् भ्रमिते भृशं भ्रमत्यर्थभूमिका तत्र ।
रथकाभ्रमरकयुक्ता परस्परं चक्रयन्त्रेण ॥186 ॥

stambhe'va(dha)stād bhramite bhṛśaṁ
bhramatyarthabhūmikā tatra |
rathakābhramarakayuktā parasparaṁ chakrayantreṇa ॥186 ॥

वसन्तरथिकाभ्रमे समधिरूढवाराङ्गना
परिभ्रमणसम्भृताभ्यधिकविभ्रमं भूपतिः ।
करोति नयनोत्सवस्त्रि(वं त्रि) दशधाम्नि
यत्कीर्तिनं वसन्तसमये भवत्यमलकीर्त्तिधामैव सः ॥187 ॥

vasantarathikābhrame samadhirūḍhavārāṅganā
paribhramaṇasambhṛtābhyadhikavibhramaṁ bhūpatiḥ |
karoti nayanotsavastri(vaṁ tri) daśadhāmni
yatkīrtanaṁ vasantasamaye bhavatyamalakīrttidhāmaiva
saḥ ॥187 ॥

In the second, the *Madanotsava*, there is no dugout
or underground construction; the storey on the main
post provides only four seats, and a man standing below
operates the machine (SS. 31.188-94).

आरोप्य स्थिरमेकं स्तम्भं भूमीगृहादिरहितमथ ।
हस्तचतुष्कोच्छ्राया कार्योपरि भूमिका चास्य ॥188 ॥

āropya sthiramekaṁ stambhaṁ bhūmīgṛhādirahitamatha |
hastachatuṣkochchhrāyā kāryopari bhūmikā chāsya ॥188 ॥

मध्ये भ्रमरकयुक्तं शेषं पूर्ववदिहाचरेदखिलम् ।
पुष्पकमिपि च स्तम्भे शिथिलं कलशोच्छ्रितं कुर्यात् ॥ 189 ॥

madhye bhramarakayuktaṁ śeṣaṁ
pūrvavadihācharedakhilam |
puṣpakamipi cha stambhe śithilaṁ kalaśochchhritaṁ
kuryāt ॥ 189 ॥

तस्योपरि च ग्रीवा चतुरासनसंयुता विधातव्या ।
घण्टास्तम्भौ कार्यौ स्तम्भेन महाबलौ तत्र ॥190 ॥

tasyopari cha grīvā chaturāsanasaṁyutā vidhātavyā |
ghaṇṭāstambhau kāryau stambhena mahābalau tatra ||190||

एवं पुष्पकभूमिकान्तरलस्थायी निगूढो
जनो यावद् भ्रामकयन्त्रचक्रनिकरं सम्यक् क्रमाच्चलयेत्।
तावत् ता रथिकासना मृगदृशस्तत्र स्थिताः
पुष्पके कामावासकुतूहलार्पितदृशो भ्राम्यन्ति सर्वा अपि ॥191॥

ēvaṁ puṣpakabhūmikāntaralasthāyī nigūḍho
jano yāvad bhrāmakayantrachakranikaraṁ samyak
kramāchchalayet |
tāvat tā rathikāsanā mṛgadṛśastatra sthitāḥ
puṣpake kāmāvāsakutūhalārpitadṛśāo bhrāmyanti sarvā
api ||191||

अथ कोणगतान् स्तम्भांश्चतुरो विनिवेशयद क्रजून् सुदृढान्।
सुश्लिष्टपीठसंस्थान् समान्तरान् मेदिनीवशतः ॥192॥।

atha koṇagatān stambhāṁśchaturo viniveśayada krajūn
sudṛḍhān |
suśliṣṭapīṭhasaṁsthān samāntarān medinīvaśataḥ ||192|| |

तेषामुपरि लता(तला) न्तरसंयुक्ता भूमिका विधातव्या।
रथिकास्तत्र चतस्रो जायन्ते पूर्ववद् दिक्स्थाः ॥193॥

teṣāmupari latā(tala) ntarasaṁyuktā bhūmikā vidhātavyā |
rathikāstatra chatasro jāyante pūrvavad diksthāḥ ||193||

तदुपरि तथार्धभूमिः कार्या सुश्लिष्टदारुसन्धाना।
मध्यभ्ररकयुक्ता सरूपका मत्तवारणयुता च ॥194॥

tadupari tathārdhabhūmiḥ kāryā suśliṣṭadārusandhānā |
madhyabhrarakayuktā sarūpakā mattavāraṇayutā cha ||194||

In the third, the *Vasanta-tilakā*, two storeys are to be constructed, the second one with much decoration; the mechanism is fitted in the first floor and by the action of gear wheel upon gear wheel, the top floor revolves (195-200).

भूषायमाणमुपयाति न विस्मयत्वं(यं कः) ॥195॥

bhūṣāyamāṇamupayāti na vismayatvaṁ(yaṁ kaḥ) ॥195॥

प्रविधाय रङ्गभूमिं प्रथमां शास्त्रान्तराधरस्यार्थे ।
चतुरश्रा रूपवती सचतुर्भद्रा विधेया भूः ॥196॥

*pravidhāya raṅgabhūmiṁ prathamāṁ
śāstrāntarādharasyārthe ।
chaturaśrā rūpavatī sachaturbhadrā vidheyā bhūḥ ॥196॥*

प्रतिकोणमागता(ता) स्या भद्रेषु भवन्ति संयता भ्रमराः ।
अत उपरिष्टाद् भूम्या भ्रमराश्चाष्टासनाः कार्याः ॥197॥

*pratikoṇamāgatā(tā) syā bhadreṣu bhavanti saṁyatā
bhramarāḥ ।
ata upariṣṭād bhūmyā bhramarāśchāṣṭāsanāḥ kāryāḥ ॥197॥*

रेखाः शुद्धाः कार्या बहिरन्तश्चित्रिताश्चान्याः ।
पीठेषु मध्यग(सं) स्थास्ततोऽपरा भूमिकाः कार्याः ॥198॥

*rekhāḥ śuddhāḥ kāryā bahirantaśchitritāśchānyāḥ ।
pīṭheṣu madhyaga(saṁ) sthāstato'parā bhūmikāḥ
kāryāḥ ॥198॥*

पीठस्य मध्यसंस्थैरन्योन्याराळियोजितैश्चक्रैः ।
सर्वे वेगाद् भ्राम्यन्ति सान्तन(रा) विभ्रमे भ्रमरारः ॥199॥

*pīṭhasya madhyasaṁsthairanyonyārāḷiyojitaiśchakraiḥ ।
sarve vegād bhrāmyanti sāntana(rā) vibhrame
bhramarāraḥ ॥199॥*

दोलासनो विहितवारवधूकृ(भृ)तातिचित्रेण यस्त्रिदशधामसु विभ्रमेण ।
पृथ्वीपतिमुदमुपैति समुल्लसन्ती कीर्त्तिर्न माति भुवनत्रितयेऽपि तस्य ॥200॥

*dolāsano vihitavāravadhūkṛ(bhṛ)tā-
tichitreṇa yastridaśadhāmasu vibhrameṇa ।
pṛthvīpatimudamupaiti samullasantī
kīrttirna māti bhuvanatritaye'pi tasya ॥200॥*

The   fourth, *'Vibhrāmaka',* provides   increased
accommodation and motion variety. At the base here is a

solid platform and a square structure with mechanism; over this is a floor with eight seats, and above these, another round of seats; spoked wheels link up the whole erection; the speciality here is that each floor has its different movements, creating, as the name implies, a complex of circular movements (SS. 31.201-208).

काणैः शेषैस्तसिंमश्चतुरश्रं कल्पयेद् भद्रम् ॥201॥

kāṇaiḥ śeṣaistasiṁmaśchaturaśraṁ kalpayed bhadram ॥201॥

तद्द्विगुणमूर्ध्वमेतस्य भूमिकाभागसङ्ख्यया कार्यम् ।
तत्राद्यंशचतुष्केण भूमिका स्यात् समुच्छ्रयतः ॥202॥

taddviguṇamūrdhvametasya bhūmikābhāgasaṅkhyayā kāryam /
tatrādyaṁśachatuṣkeṇa bhūmikā syāt samuchchhrayataḥ ॥202॥

तत्राष्टषड्चतुर्भागवर्जिता भूमिका उपर्युपरि ।
क्रमशो भवन्त्यथैवं ताः स्युस्तिस्रोऽर्धसंयुक्ताः ॥203॥

tatrāṣṭaṣaṭchaturbhāgavarjitā bhūmikā uparyupari /
kramaśo bhavantyathaivaṁ tāḥ syustisro'rdhasaṁyuktāḥ ॥203॥

शेषांशोच्छ्रययुक्ता घण्टा चतुरश्रकायता कार्या ।
त्रिचतुर्भूम्यौ कार्ये सषड्चतुर्भागविस्तारे ॥204॥

śeṣāṁśochchhrayayuktā ghaṇṭā chaturaśrakāyatā kāryā /
trichaturbhūmyau kārye saṣaṭchaturbhāgavistāre ॥204॥

रङ्गस्यादाद्यभूवि द्वितीयभुवि कोणगास्तथा रथिकाः ॥
स्युर्भद्रकृतियुक्ता दोला अपि तत्र रमणीयाः ॥205॥

raṅgasyādādyabhūvi dvitīyabhuvi koṇagāstathā rathikāḥ ॥
syurbhadrakṛtiyuktā dolā api tatra ramaṇīyāḥ ॥205॥

रथिकास्तृतीयभूमौ कार्या भद्रेषु चातिरमणीयाः ।
कोणेष्वथासनान्यर्धवास्तुकेऽपि भ्रमः कार्यः ॥206॥

rathikāstṛtīyabhūmau kāryā bhadreṣu chātiramaṇīyāḥ /
koṇeṣvathāsanānyardhavāstuke'pi bhramaḥ kāryaḥ //206 //

दोलारथिक चतुरासने भ्रमोष्टासनो भवेत् तत्र ।
आसनमिह तत् कथितं युवतेः स्थानं यदेकं स्यात् ॥207 ॥

dolārathika chaturāsane bhramoṣṭāsano bhavet tatra /
āsanamiha tat kathitaṁ yuvateḥ sthānaṁ yadekaṁ syāt //207

निखिलान्यपि भ्रमणसंमुखं तानि बिभ्रति भ्रमणम् ।
यत्रासनानि स इह भ्रम इत्युक्तोऽपराधिका ॥208 ॥

nikhilānyapi bhramaṇasaṁmukhaṁ tāni bibhrati
bhramaṇam /
yatrāsanāni sa iha bhrama ityukto'parādhikā //208 //

The last, *Tripura,* increases the tiers by one, justifying its name of three cities in the air, each higher floor being of smaller dimensions; a large number of connecting links, small wheels, and steps leading from one tier to the other are mentioned (SS. 31.209-218).

यष्टेरूर्ध्वमधस्ताद् भ्रमस्य चक्रं नियोजयेदेकम् ।
लघुचक्राणि च तद्वन्नियोजयेदासनेष्वत्र ॥209 ॥

yaṣṭerūrdhvamadhastād bhramasya chakraṁ
niyojayedekam /
laghuchakrāṇi cha tadvanniyojayedāsaneṣvatra //209 //

लघुचक्रारकवृत्ते संलग्नाः कीलका दृढाः कार्याः ।
तुल्यान्तराः समस्ताः प्रलघु(क) चक्रारवृन्तगताः ॥210 ॥

laghuchakrārakavṛtte saṁlagnāḥ kīlakā dṛdhāḥ kāryāḥ /
tulyāntarāḥ samastāḥ pralaghu(ka)
chakrāravṛntagatāḥ //210 //

रथिकाशिखाग्रचक्रं भ्रमचक्रारक विनियोजितं कार्यम् ।
यष्टिचुष्टयमसिंमस्तिर्यक् चक्रद्वयोपेतम् ॥211 ॥

rathikāśikhāgrachakraṁ bhramachakrāraka viniyojitaṁ
kāryam /

yaṣṭichuṣṭayamasiṁmastiryak chakradvayopetam ||211||

ऊर्ध्वं द्वितीयभूमेस्तृतीयभूमेरथान्तरे कुर्यात्।
नियतं रथिकायष्टिभ्रमसंलग्नानि यन्त्राणि ||212||

ūrdhvaṁ dvitīyabhūmestṛtīyabhūmerathāntare kuryāt |
niyataṁ rathikāyaṣṭibhramasaṁlagnāni yantrāṇi ||212||

आसानाधारयष्टीनां रथिकाचक्रयोजितान्।
अधः समान्तरान् कुर्याच्चतुरः परिवर्तकान् ||213||

āsānādhārayaṣṭīnāṁ rathikāchakrayojitān |
adhaḥ samāntarān kuryāchchaturaḥ parivartakān ||213||

त(द्व)द् द्वितीयभूमीदोलागर्भे समान्तरे यष्टी।
लग्ने तथैकचक्रे याम्योत्तरचक्रयोर्न्यस्येत् ||214||

ta(dva)d dvitīyabhūmīdolāgarbhe samāntare yaṣṭī |
lagne tathaikachakre yāmyottarachakrayornyasyet ||214||

तद्वदधो भूकोणगरथिकाचूडाग्रचक्रसंसक्ताः।
यष्टीस्ततश्चतस्रो द्विचक्रका इतरचक्रयोर्न्यस्येत् ||215

tadvadadho bhūkoṇagarathikāchūḍāgrachakrasaṁsaktāḥ |
yaṣṭīstataśchatasro dvichakrakā itarachakrayornyasyet ||215

प्रान्तचक्रद्वये कोणरथिकाचक्रयोजिता।
दोलागर्भगता यष्टिस्तिर्यक् कार्यापरापरा ||216||

prāntachakradvaye koṇarathikāchakrayojitā |
dolāgarbhagatā yaṣṭistiryak kāryāparāparā ||216||

पूर्वे भद्रे द्वारं कुर्यात् सोपानराजितमधस्तात्॥
गर्भात् परिश्चमभागे निवेशयेद् देवतादोलाम् ||217||

pūrve bhadre dvāraṁ kuryāt sopānarājitamadhastāt ||
garbhāt pariśchamabhāge niveśayed devatādolām ||217||

अन्योन्यं चक्रभ्रममिच्छामुक्तिं विधानतः सम्यक्।
ज्ञात्वा प्रयोजनीयं शीघ्रवहं मन्दवहनं वा ||218||

anyonyaṁ chakrabhramamichchhāmuktiṁ vidhānataḥ samyak |
jñātvā prayojanīyaṁ śīghravahaṁ mandavahanaṁ vā ||218||

The *Samarāṅgaṇa Sūtradhāra* (Chapter 31) of Rājā Bhoja (336-392 A.D.) describes various attributes of machines as the proper union of effort and result, good contact, smoothness, requiring no attention, sustaining, lightness, and noiselessness, when sound is required its predominance, not loosening or clogging, good contact in all moving parts, no break in the action, attainment of results as desired, adherence to rhythm and timing, showing results at the moment desired, returning to normalcy at other times, not bulging and staying in shape, strength, softness and durability.[64] A few examples of machines invented in ancient India are illustrated below.

---

64 यथावद्बीजसंयोगः सौश्लिष्ट्यं श्लक्ष्णतापि च। अलक्ष्यतानिर्वहणं लघुत्वं शब्दहीनता। शब्दे साध्ये तदाधिक्यमशैथिल्यमागाढता। वहनीषु समस्तासु सौश्लिष्ट्यं चासलद्गतिः। यथेष्टार्थकारित्वं लयतालानुगामिताः। इष्टकालेऽर्थदर्शित्वं पुनः सम्यक्तवसंवृति॥ अनुल्बणत्वं ताद्रूप्यं दार्ढ्यमसृणता तथा। चिरकालसहत्वं च यन्त्रस्यैते महागुणाः॥

*yathāvadbījasaṁyogaḥ   sauśliṣṭyaṁ   ślakṣaṇatāpi   cha /
alakṣayatānirvahaṇaṁ   laghutvaṁ   śabdahīnatā /   śabde   sādhye
tadādhikyamaśaithilyamāgāḍhatā /   vahanīṣu samastāsu sauśliṣṭyaṁ
chāsaladgatiḥ   yatheṣṭārthakāritvaṁ   layatālānugāmitāḥ /
iṣṭakāle'rthadarśitvaṁ punaḥ samyaktavasaṁvṛti // anulbaṇatvaṁ
tādrūpyaṁ   dārḍhyamasṛṇatā   tathā /   chirakālasahatvaṁ   cha
yantrasyaite mahāguṇāḥ //*

8

Marine Engineering

Indian marine engineering has a long, glorious past. Ancient Indians used the vessels in rivers and seas for commercial and maritime activities. The Indian ships would ply far and wide on different routes and seas. The *Ṛgveda,* which is the most ancient authentic work, contains a reference to a ship with one hundred paddles or ores.

तिस्रः क्षपस्त्रिरहातिव्रजद्भिर्नासत्या भुज्युमूहथुः पतङ्गैः ।
समुद्रस्य धन्वन्नार्द्रस्य पारे त्रिभी रथैः शतपद्भिः षळश्वैः ॥4॥

*tisraḥ kṣapastrirahātivrajadbhirnāsatyā bhujyumūhathuḥ
patangaiḥ |
samudrasya dhanvannārdrasya pāre tribhī rathaiḥ
śatapadbhiḥ ṣaḷaśvaiḥ ॥4॥*

(Tribhir-rathaiḥ) Three kinds of transport- surface, marine, and air is required *(pāre)* for moving/travelling on earth *(ārdrasya samudrasya)* sea full of water and *(dhanvanaḥ)* space. These three means of transport should be *(ati-vrajadbhiḥ)* fast-moving as if *(śatapadbhiḥ)* propelled by 100 wheels (in case of surface transport), 100 paddles (in case of a ship) and 100 propellers (in case of an airplane) *(patangaiḥ)* fitted with one or more engines and *(ṣaḍaśvaiḥ)* and six powers of acceleration (gears). These vehicles may help you reach your destination within *(tisraḥ)* three *(kṣapaḥ)* nights or *(triḥ)* three *(aha)* days.

अनारम्भणे तदवीरयेथामनास्थाने अग्रभणे समुद्रे ।
यदश्विना ऊहथुर्भुज्युमस्तं शतारित्रां नावमातस्थिवांसम् ॥5 ॥

anārambhaṇe tadavīrayethāmanāsthāne agrabhaṇe
samudre /
yadaśvinā ūhathurbhujyumastam śatāritrām
nāvamātasthivāmsam //5 //

(Aśvinau) O merchants and traders you (avīrayethā) venture into (samudre) in the ocean of water and in the space (tad) where there is (anārambhaṇe) no support, (anāsthāne) nothing to rest upon, (agrabhaṇe) nothing to cling to (ātasthivānsam) by boarding the (nāvaŠ) the ship (śatāritrā) (astam) steered by hundreds of propellers/paddles (ūhathuḥ) for carrying the (bhujyum) products or items of use from one country to another.

There is also evidence of commercial shipping earlier to the ruins of Ur and Yul Bagas- the first king of united Babylonia ruled in Ur in 3000 B.C. The kind of teak was identified as belonging to the Malabar coast. The same must have been sent by sea.

Rāmāyaṇa (Ayodhyā kāṇḍa) has the description of the warship as:

नावां शतानां पंचानां कैवर्तानां शतं शतम् ।
सनद्धानां तथा यूनान्तिष्ठक्वित्यभ्यचोदयत् ॥

nāvāṁ śatānāṁ paṁchānāṁ kaivartānāṁ śataṁ śatam /
sanaddhānāṁ tathā yūnāntiṣṭhakvityabhyachodayat //

"That is, hundreds of sailors are sailing five hundred ships carrying hundreds of armed soldiers."

The Construction of the overpass on the sea to reach the island Śrilaṅkā is another example of early navigation and sea exploration from India to other regions.

Similarly, there is a description of the mechanized ship in *Mahābhārata as-*

सर्ववातसहां नावं यन्त्रयुक्तां पताकिनीम् ॥

sarvavātasahāṁ nāvaṁ yantrayuktāṁ patākinīm ॥

"That is a ship propelled by machine, and a flag can work in all sorts of winds."

During the period of Maurya (15th century B.C.), Gupta (3rd Century BC), and Harsh Vardhan (487 BC), sea voyages were made to Malaya Peninsula, and the colonization of Annam Cambodia and Java by Indians were possible only through maritime shipping activities.

Also, in Kautilya's *Arthashastra* (15th century B.C.), we find information on the complete arrangements of boats maintained by the navy and the state. It also contains information on the duties of the various personnel on a ship. For example, the Nāvādhyakṣa is the superintendent of the ship, Niyāmaka is the Steerman, and Dātragrahaka is the holder of the needle or the compass. Differences in ships are also described regarding the cabins' location and the ship's purpose.

All these brisk shipping and navigation activities were only possible with a sound knowledge of the technology of fabricating boats and ships. There must have been ship designers, artisans, and other experts relating to the art and science of shipbuilding. In this context, a great ship design engineer named Bhoja Narpati (336-392 A.D.) is worth mentioning. There are two persons in the name of Bhoja (336-392 A.D.). One was the king of Dhar of Malwa Kingdom, and another was Bhoja Narpati (336-392 A.D.), who appeared to be an expert in designing ships and boats. Yukti kalpa taru is the work of Bhoja Narpati (336-392 A.D.).

The Yuktikalpataru (4th century A.D.) deals with the subject of shipbuilding elaborately in two chapters entitled Niṣpada-yānoddeśa and Jaghanya Jalayānani under the following heads:

(a) seasons or periods suitable for shipbuilding;

(b) varieties of woods best suited for the construction of ships; (c). Bhoja's injunction regarding tying an iron nail to a sea-going vessel; c). classification of ships-river-going or ordinary (sāmānya) and sea-going or special (viśeṣa); (e) names and measurements of ordinary (sāmānya) type of vessel; (f) two types of special (viśeṣa) ship-dīrgha (according to length) and unnata (according to height); (g) names and measurements of dirgha type of ship; (h) Bhoja's view about dīrgha type of ship; (i) names and measurements of unnata type of ship; (j) Bhoja's opinion about unnata type of ship; (k) painting of ships; (l). decoration of ship; (m) ships with cabins; (n) characteristics of the royal ship according to Bhoja; (o) despicable water vessels (Jaghanya jalayanani).

According to *Yuktikalpataru* (4th century A.D.), ship or boat comes in the category of un-wheeled vehicles.

Classification of Vessels

The vessels were classified into the following two main categories, viz. Sāmānya and Viśeṣa. (*Yuktikalpataru*, 89)

अथ लक्षणानि । सामान्यंच विशेषश्च नौकायाः ॥ 89 ॥

atha lakṣaṇāni I sāmānyaṁcha viśeṣaścha naukāyāḥ ॥ 89 ॥

Sāmānya types of Vessels: Sāmānya vessels were simple and ordinary vessels mostly plied in rivers.

Sāmānya vessels were classified into ten types: kṣudra (smallest), madhyama (moderate), bhima (formidable), capala (swift moving), paṭala (with covering), abhaya (protecting from dangers) dirgha (tall), patraputa (like that of the cup made of a folded leaf), garbhara (with inner compartments) and manthara (curved). (*Yuktikalpa Taru*, 92)

Kṣudra: The vessel whose length is one rājahasta (16 cubit or 32 ft.), breadth one-fourth, and height is just the same (of the breadth) is called kṣudra. The dimension of the kṣudra vessel was 16 x 4 x 4 cubits or 32 ft. x 8 ft. x 8 ft. (sq. ft.) (*Yuktikalpa Taru*, 90)

Madhyamā: The ship whose length is one and half of the rājahasta, the breadth half of the length and the height one-third, is called madhyama. The dimension of the kṣudra vessel was 24 x 12 x 8 cubits or 48 ft. x 24 ft. x 16 ft. (*Yuktikalpa Taru*, 91)

The respective measurements of the length of the ten ships should be known as one (rājahasta, i.e. 16 cubits or 32 ft.), an increment of the same, an increment of the same by its half, and so on alternately, and those of their height and breadth are also follow the same rule.

नौका दशकमित्युक्तं राजहस्तैरनुक्रमम् ।
एकैकवृद्धै सार्धैश्च विजानीयात् द्वयं द्वयम् ॥ *(Yuktikalpa Taru, 93)*

naukā daśakamityuktaṁ rājahastairanukramam /
ēkaikavṛddhai sārdhaiścha vijānīyāt dvayaṁ dvayam //

(Yuktikalpa Taru, 93)

The dimensions of other vessels are illustrated below:

Name of the vessel	Length in rājahasta (ft.)	Breadth in rājahasta (ft.)	Height in rājahasta (ft.)
Kṣudra (Smallest)	16 (32)	4 (8)	4 (8)
Madhyamā	24 (48)	12 (24)	8 (16)
Bhimā (formidable)	40 (80)	20 (40)	12 (24)
Capalā (swift moving)	48 (96)	24 (48)	16 (32)
Paṭalā (with covering)	64 (128)	32 (64)	20 (40)
Abhayā (protecting from dangers)	72 (144)	36 (72)	24 (48)
Dirghā (tall)	80 (160)	40 (80)	28 (56)
Patrapuṭā (like that of cup made of folded leaf)	96 (192)	48 (96)	32 (64)
Garbharā (with inner compartments)	104 (208)	52 (104)	36 (72)
Mantharā (curved)	120 (240)	60 (120)	40 (80)

Besides mantharā, the rest are seagoing vessels, and as regards their merits, they are, briefly speaking, endurable and spacious.

Viśeṣa types of Vessels: Viśeṣa type of vessels were solid and sturdy and were meant for navigation and maritime activities in the seas. They were made of the foil of metals like copper, etc. or loadstone. They were further classified into two categories viz. Dīrghā as per length and Unnatā as per height.

लौहाताम्रादिपत्रेण कान्तलौहेन वा तथा ।
दीर्घा चैवोन्नता चेति विशेषद्विविधा भिदा ॥ (*Yuktikalpa Taru*, 96)

lauhātāmrādipatreṇa kāntalauhena vā tathā /
dīrghā chaivonnatā cheti viśeṣadvividhā bhidā //

(*Yuktikalpa Taru, 96*)

Dirghā types of Vessels: The vessel whose length is two rājahasta breadth one-eighth (of its length) and height one-tenth (of its length) is called dīrghikā.

राजहंसद्वयायामा अष्टांशपरिणाहिनी ।
नौकेयं दीर्घिका नाम दशांकेनोन्नतापि च ॥ (*Yuktikalpa Taru*, 97)

rājahaṁsadvayāyāmā aṣṭāṁśapariṇāhinī /
naukeyaṁ dīrghikā nāma daśāṁkenonnatāpi cha //

(*Yuktikalpa Taru, 97*)

Dirghā types of vessels were further divided into ten types: dīrghikā (tall), taroṇi, lolā (moving hither and thither) gatvārā (perishable), gāminī (going and moving on), tari, jaṅghāla (running swiftly), plāvinī (flowing over), dhāriṇī (power of possessing), veginī (having velocity).

दीर्घिकातरणिर्लोला गल्वरा गामिनी तरिः ।
जंघाला प्लाविनी चैव धारिणी वेगिनी तथा ॥ (*Yuktikalpa Taru*, 98)

dīrghikātaraṇirlolā gatvarā gāminī tariḥ /
jaṁghālā plāvinī chaiva dhāriṇāī veginī tathā //

(*Yuktikalpa Taru, 98*)

Each of whose length is increased by one rajahasta in arithmetic progression and each of whose height and breadth is one-tenth and one-eighth of the same (length) respectively.

राजहस्तैकैक वृद्धा द्रया नौका नामानि वै दश ।

उन्नति परिणाहश्च दशाष्टमशंमितौ क्रमात् ॥ (*Yuktikalpa Taru*, 99)

rājahastaikaika vṛddhā dvayā naukā nāmāni vai daśa /
unnati pariṇāhaścha daśāṣṭamaśaṁmitau kramāt //

(*Yuktikalpa Taru, 99*)

The dimensions of various dīrghikā vessels are illustrated below:

Name of the vessel	Length in rājahasta (ft.)	Breadth in rājahasta (ft.)	Height in rājahasta (ft.)
Dīrghikā	32 (64)	4 (8)	3.2 (6.4)
Taraṇī	48 (96)	6 (12)	4.8 (9.6)
Lolā	64 (128)	8 (16)	6.4 (12.8)
Gatvārā	80 (160)	10 (20)	8 (16)
Gāminī	96 (192)	12 (24)	9.6 (19.2)
Tari	112 (224)	14 (28)	11.2 (22.4)
Jaṅghāla	128 (256)	16 (32)	12.8 (25.6)
Plāvinī	144 (288)	18 (36)	14.4 (28.8)
Dhāriṇī	160 (320)	20 (40)	16 (32)
Veginī	176 (352)	22 (44)	17.6 (35.2)

We have seen above that Dīrgha types of vessels had ten varieties. The largest size is 176 x 22 x 17 cubits or 352 ft. x 44 ft. x 35.2 ft. (sq. ft.)

Bhoja also says (about dīrghikā type of vessels) that though the length of the ship may be made according to one's desire, it should be restricted to the numbers of the two hands, eight Vasus, four Vedas and nine grahas (planet).

भोजोऽपि

नौका दीर्घं यथेच्छं स्यात् तत्रैतानि विवर्जयेत् ।

हस्तसंख्या परित्याज्या वसुवेदग्रहोत्तरे ॥ (*Yuktikalpa Taru*, 102)

bhojo' pi

naukā dīrghaṁ yathechchhaṁ syāt tatraitāni vivarjayet /

hastasaṁkhyā parityājyā vasuvedagrahottare //

(Yuktikalpa Taru, 102)

Then comes the ship according to height (unnatā). Ships with two rājahastas in length the breadth, and the height.

राजहस्तद्वयमिता तावत् प्रसरनोन्नता

इयोमूर्घ्वाभिधा नौका क्षेमाय पृथिवीभूजाम् ॥

rājahastadvayamitā tāvat prasaranonnatā

iyomūrghvābhidhā naukā kṣemāya pṛthivībhūjām //

Unnatā types of vessels: Unnatā types of vessels are divided into five types: urdhvā (elevated), anurdhvā (non-elevated), svarṇamukhi (good-faced), garvinī (power of being filled with), and mantharā (slow-going). They are the types of the five ships with each of their length increased by one rājahasta in arithmetic progression.

Name of the vessel	Length in rājahasta (ft.)	Breadth in rājahasta (ft.)	Height in rājahasta (ft.)
Urdhvā	32 (64)	4 (8)	3.2 (6.4)
Anurdhvā	48 (96)	20 (40)	19.2 (38.4)
Svarṇamukhi	64 (128)	36 (72)	38.2 (76.4)
Garvinī	80 (160)	52 (104)	54.2 (108.4)
Mantharā	96 (192)	68 (136)	70.2 (142.4)

Thus, we see that the Unnata class of vessels was of an advanced type. The most significant size among them was 96 x 68 x 70.2 cubits. (192 ft. x 136 ft. x 142.4 ft. (sq. ft.)

Painting and Design of Vessels

Vessels were painted depending upon the number of masts they had. For example-

The vessels with one mast were painted blue.

The vessels with two masts were painted yellow.

The vessels with three masts were painted red.

The vessels with four masts were painted white.

The shapes of the fronts of ships were designed into eight types: lion, tiger, duck or parrot, buffalo, serpent, elephant, tiger, bird, frog, and man.

The picture of celestial bodies such as those of the sun pitcher, mirror, the moon, swan, peacock, parrot, lion, elephant, serpent, tiger, two bees, Indra, etc., were depicted in the body of the vessels after smearing the same with fourth (white), third (red), second (yellow) and first (blue) colours in due proportion one after another. The verses in Yuktikalpa Taru read under:

ब्रह्मादिभिः परिन्यस्य नौका चित्रणकर्मणि ।
चतुश्रृंगा त्रिश्रृंगा द्विश्रृंग चैकश्रृंगिणी ॥ 110 ॥

brahmādibhiḥ parinyasya naukā chitraṇakarmaṇi |
chatuśṛṁgā triśṛgā dviśṛṁga chaikaśṛṁgiṇī || 110 ||

सितरक्तापीतानिलवर्णान् दद्यात् यथाक्रमम् ।
केसरी महिषो नागो द्विरदो व्याघ्रेव च ॥ 111 ॥

sitaraktāpītānilavarṇān dadyāt yathākramam |
kesarī mahiṣo nāgo dvirado vyāghreva cha || 111 ||

पक्षीभेको मनुष्यश्च एतेषां वदनाष्टकम् ।
नावं मुखे परिन्यस्य आदित्यादि-दशा-भुवम ॥ 112 ॥

pakṣībheko manuṣyaścha ēteṣāṁ vadanāṣṭakam |
nāvaṁ mukhe parinyasya ādityādi-daśā-bhuvama || 112 ||

कलसो दर्पणश्चन्द्रस्त्रैदशानं महीभुजाम् ।
हंसः केकी शुकः सिंहो गजोऽहि व्याघ्रषेतपदौ ॥ 113 ॥

kalaso darpaṇaśchandrastraidaśānaṁ mahībhujām |
haṁsaḥ kekī śukaḥ siṁho gajo'hi vyāghraṣetapadau || 113

आदित्यादिश्च जाता नौकोपरि परिन्यसेत् ।
चतुस्त्रित्वैक विमिता चतुवर्णा यथाक्रमम् ॥ 114 ॥

ādityādiścha jātā naukopari parinyaset |
chatustritvaika vimitā chatuvarṇā yathākramam || 114 ||

Decoration of the Ship

The ship, whose four-sides arc wrapped with thin sheets of metal and whose length is one-fourth of one hundred (twenty-five), whose breadth is half of its length (twelve and a half) and whose height is half (six and one-fourth) of its breadth is called by the name Kamalā. The eight colours of the kind of Avājñāsika cloth should be known as being composed of white, red, spotted yellow and black colours, mixing three in each case. The ships studded with pearls resembled the umbrella of navadaṇḍa type. The number of pearls bedecking the ship should be two, four, and six, respectively. Honest people call the gold necklace of the ship the garland of victory. The verses in Yuktikalpataru read under:

अच्छादनं चतुष्पते कमला नाम कथ्यते ।
तत् संख्याशतपदे मे तदर्धार्धमिवापरात् ॥ 115 ॥

achchhādanaṁ chatuṣpate kamalā nāma kathyate |
tat saṁkhyāśatapade me tadardhārdhamivāparāt || 115 ||

शुक्लरक्तोथ चित्रश्च पीतः कृष्णस्त्रिभिस्त्रिभिः ।
अवज्ञासिकसंज्ञानां वस्त्र वर्णाष्टकं विदुः ॥ 116 ॥

śuklaraktotha chitraścha pītaḥ kṛṣṇastribhistribhiḥ /
avajñāsikasaṁjñānāṁ vastra varṇāṣṭakaṁ viduḥ // 116 //

नौकासु मणिविन्यासो विज्ञेयो नवदण्डवत् ।
मुक्तास्तवकैर्युक्ता नौका स्यात् सर्वतोभद्रा ॥ 117 ॥

naukāsu maṇivinyāso vijñeyo navadaṇḍavat /
muktāstavakairyuktā naukā syāt sarvatobhadrā // 117 //

तत् संख्याचेदथ रसवेदद्द्वय सम्मिता क्रमशः ।
कनकादीनां मनोज्यमालेति गद्यते सद्भिः ॥ 118 ॥

tat saṁkhyāchedatha rasavedadvaya sammitā kramaśaḥ /
kanakādīnāṁ manojyamāleti gadyate sadbhiḥ // 118 //

Vessels with Cabins

The vessels were also equipped with cabins. For the sake of cabins, vessels were classified into four classes: Brahma, kṣatra, vaiśya, and śudra. They were further classified into two types: with cabins and without cabins.

ब्रह्मक्षत्रे द्वितये एकैके वैश्यशूद्रयोनी ।
निर्गृहं सगृहं वाथ तत्सर्वं द्विविधं भवेत् ॥ 119 ॥

brahmakṣatre dvitaye ēkaike vaiśyaśūdrayonī /
nirgṛhaṁ sagṛhaṁ vātha tatsarvaṁ dvividhaṁ bhavet // 119

All the ships mentioned above fall in the category of ships without cabins. Ships with cabins were divided into three categories—sarvamandirā, madhyamandirā and agramandirā. The location of cabins would depend upon the purpose for which they were to be used.

सगृहा त्रिविधा प्रोक्ता सर्वमध्याग्रमन्दिरा ॥ 120 ॥
sagṛhā trividhā proktā sarvamadhyāgramandirā // 120 //

The vessels with large capacities used for the transport

of royal treasure, horses, and women had cabins extending from one end to the other and were called sarvamandirā.

सर्वतोमन्दिरं यत्र सा ज्ञेया सर्वमन्दिरा ।
राज्ञां कोशाश्वनारीणां यानमत्र प्रशस्यते ॥ 121 ॥

sarvatomandiram yatra sā jñeyā sarvamandirā ǀ
rājñām kośāśvanārīṇām yānamatra praśasyate ǁ 121 ǁ

The vessels are suitable in the rainy season and meant for pleasure trips for kings who used to have cabins in the middle part. They are called madhyamandirā.

मध्यतो मन्दिरं यत्र सा ज्ञेया मध्यमन्दिरा ।
राज्ञां विलासयात्रादि वर्षासु च प्रशस्यते ॥ 122 ॥

madhyato mandiram yatra sā jñeyā madhyamandirā ǀ
rājñām vilāsayātrādi varṣāsu cha praśasyate ǁ 122 ǁ

The vessels, convenient for dry seasons and meant for naval warfare and long journeys, had cabins towards the front side. Such vessels are called agramandirā.

अग्रतो मन्दिरं यत्र सा ज्ञेयाग्रमन्दिरा ।
चिरप्रवास यात्रायां रणे काल घनात्यये ॥ 123 ॥

agrato mandiram yatra sā jñeyāgramandirā ǀ
chirapravāsa yātrāyām raṇe kāla ghanātyaye ǁ 123 ǁ

The ship, having a cabin less than half of its length, becomes swift in speed.

मन्दिरामानं नौका प्रसरत एवार्ध भागतो न्यूनाम् ॥ 123

mandirāmānam naukā prasarata ēvārdha bhāgato nyūnām ǁ

Royal Vessels

According to Bhoja, there was another category of vessels except the four mentioned above. This fifth category of vessels was known as the royal ship. It had an oval shape, and the heights of its cabins were six, eight, nine, ten, twelve, and five rājahastas respectively.

दीर्घवृत्तवसु(8)षट्(6)दिवाकरानेक(12)दिङ्(4)नव(9)मिता
यथाक्रमम्। राजपंचभुजसम्मितोन्नतिर्मन्दिरे त्रिगते महीभुजाम्॥ 124 ॥

dīrghavṛttavasu(8)ṣaṭ(6)divākarāneka(12)diṅ(4)nava(9)
mitā yathākramam /
rājapaṁchabhujasammitonnatirmandire trigate
mahībhujām || 124 ||

The picture of the celestial bodies, such as those of the suns, etc., were depicted in the royal ship with gold, silver, or copper metalwork, and those of the flags and pitchers, etc., were studded with pearls resembling an umbrella of navadaṇḍa.

भास्करादिक.दशभुवां पुनर्धातु निर्णयमत्रपूर्ववत्।
पताकाकलशादिनां निर्णयो नवदण्डवत्॥ 125 ॥

bhāskarādika.daśabhuvāṁ punardhātu
nirṇayamatrapūrvavat /
patākākalaśādināṁ nirṇayo navadaṇḍavat || 125 ||

The cabins of the royal ship were made either of wood, conducive to prosperity and happiness, or of metal for pleasure.

काष्ठजं धातुजं चेति मन्दिरं द्विविधं भवेत्।
काष्ठजं सुखसम्पत्त्यै विलासे धातुजं मतम्॥ 126 ॥

kāṣṭhajaṁ dhātujaṁ cheti mandiraṁ dvividhaṁ
bhavet /
kāṣṭhajaṁ sukhasampattyai vilāse dhātujaṁ matam ||

They were decorated with bedsteads, seats, curved canopies, and other furniture.

अथ शय्यासनादिनां मन्थरोल्लोचयोरपि ।
अन्येषां चैव मुनिभिः निर्णयः पूर्ववन्मतः ॥ 127 ॥

atha śayyāsanādināṁ mantharollochayorapi ǀ
anyeṣāṁ chaiva munibhiḥ nirṇayaḥ pūrvavanmatah ǁ

Lightness, hardness, mobility hollowness, and good-proportion-these are known to be the attributes of the ships.

लघुता दृढता चैव गामिताज्छिद्रता तथा ।
समतेति गुणोद्देशो नौकायां सम्प्रकाशितः ।।

laghutā dṛḍhatā chaiva gāmitājchhidratā tathā ǀ
samateti guṇoddeśo naukāyāṁ samprakāśitaḥ ǀ

The king who builds his ship after considering all these enjoys happiness throughout his whole life and gains victory in war.

एवं विचिन्त्य यो राजा नौकायानानि करोति च ।
सः चिरं सुखमाप्नोति विजयं समरेश्रियम् ॥ 130 ॥

ēvaṁ vichintya yo rājā naukāyānāni karoti cha ǀ
saḥ chiraṁ sukhamāpnoti vijayaṁ samareśriyam ǁ130

The king who builds the ship otherwise out of his ignorance, his reputation, power, and wealth are all destroyed.

यो अज्ञानादन्यथा यानं नौकानं कुरुते नृपः ।
तस्यैतानि विनश्यन्ति यशो वीर्यं बलं धनम् ॥ 131 ॥

yo ajñānādanyathā yānaṁ naukānaṁ kurute nṛpaḥ ǀ
tasyaitāni vinaśyanti yaśo vīryaṁ balaṁ dhanam ǁ 131 ǁ

Material used

The vessel's main body used to be constructed with

wood qualified for maritime purposes and having immunity from rot or disintegration into seawater.

"In the Vṛkṣāyurveda (the Science of Plantlife) it is stated that there are four types of trees which have their four varieties of woods altogether; such as brahma-jāti-the wood which is light, soft and can be quickly joined; kṣatrajāti-the weed which is light, stiff and can not be joined; the wood which is soft and heavy and the wood which is hard and heavy are respectively called vaiśya and śudrajāti. The wood of mixed class consists of two (separate) properties.

वृक्षायुर्वेद गदिता वृक्षजातिश्चतुर्विधा ।
समासेन गदितं तेषां काष्ठं चतुर्विधम् तद्यथा ॥ 83 ॥

vṛkṣāyurveda gaditā vṛkṣajātiśchaturvidhā /
samāsena gaditaṁ teṣāṁ kāṣṭhaṁ chaturvidham tadyathā //

लघु यत्कोमलं काष्ठसुघटं ब्रह्मजाति तत् ।
दृढांगं लघु यत्काष्ठमघटं क्षत्रजाति तत् ॥ 84 ॥

laghu yatkomalaṁ kāṣṭhaṁsughaṭam brahmajāti tat /
dṛḍhāṁgaṁ laghu yatkāṣṭhamaghaṭaṁ kṣatrajāti tat // 84 //

कोमलं गुरु यत्काष्ठं वैश्य जाति तदुच्यते ।
दृढांगं गुरु यत्काष्ठं शुद्रजाति तदुच्यते ।
लक्षणद्वययोगेन द्विजातिः काष्ठसंग्रह ॥ 85 ॥

komalaṁ guru yatkāṣṭhaṁ vaiśya jāti taduchyate /
dṛḍhāṁgaṁ guru yatkāṣṭhaṁ śudrajāti taduchyate /
lakṣaṇadvayayogena dvijātiḥ kāṣṭhasaṁgraha // 85 //

The ship constructed with the kṣatriya class's wood was considered of superior quality. For passing through troubled water, the ship was made of light and hardwood.

क्षत्रिय काष्ठैर्घटित भोजमते सुखसम्पदं नौका ।

अन्ये लघुभिः शुद्रैः विद्धति जलदुष्पदे नाकाम् ॥ 86 ॥

kṣatriya kāṣṭhairghaṭita bhojamate sukhasampadam naukā /
anye laghubhiḥ śudraiḥ viddhati jaladuṣpade nākām // 86 //

The ships of two kinds of (mixed) wood were considered poor quality. They did not last for a long time; they rotted, split and sank in the water".

विभिन्नजातिद्वय काष्ठजाता न श्रेयसे नापि सुखाय नौका ।
नैषा चिरं तिष्ठति पच्यते च बिभिद्यते वारिणि मज्जते च ॥ 87 ॥

vibhinnajātidvaya kāṣṭhajātā na śreyase nāpi sukhāya
naukā /
naiṣā chiram tiṣṭhati pachyate cha bibhidyate vāriṇi majjate
cha // 87 //

While manufacturing and decorating vessels, metals were also used, except iron. Mainly gold, silver, copper, and the compound of all these three were used.

धात्वादिनामतो वक्षे निर्णर्यं तरीसंश्रयम् ।
कनकं रजतं ताम्रं त्रितयं वा यथाक्रमम् ॥ 89 ॥

dhātvādināmato vakṣe nirṇaryam tarīsaṁśrayam /
kanakam rajatam tāmram tritayam vā yathākramam // 89 //

Prohibition of use of steel/iron

The use of iron or steel in the fabrication of vessels was abandoned entirely to avoid any risk with the magnetic rocks present in the seabed. We can also say that planks of the hull's bottom were held together with the help of substances other than iron to avoid this risk.

न सिन्धुगाद्यार्हति लौहबन्धनम् तल्लोहकान्तैः ह्रियते हि लौहम् ।
विपद्यते तेन जलेषु नौका गुणेन बन्धं निजगाद् भोजः ॥ 88 ॥

na sindhugādyārhati lauhabandhanam tallohakāntaiḥ
hriyate hi lauham /

vipadyate tena jaleṣu naukā guṇena bandhaṁ nijagād
bhojaḥ || 88 ||

Here, it may be informed that many European travellers in the Middle Ages have generally commented on the absence of iron nails in Indian shipping. In 1292 AD, John of Montecorvino (H. Yule, 1914-17: vol.1, p.66-67) mentions Indian Ships as "mighty frail and uncouth with no iron in them". Marco Polo (1903: vol.1, p.108) remarked, "They have no iron to make nails of." Vasco Da Gama (p. 124 and n.4) explained, "Some of these vessels are without any nails or iron for they have to pass over the lodestone."

Low-quality vessels

Deep research was done on the matter of quality and durability of vessels. Some vessels were identified as low quality. Among such vessels were included water vessels such as vessels other than ships (aforesaid) and those of the Western people, which are said to have many shapes; they are:

नौकायन्तो जले यानं जघन्यमिति गद्यते ।
तद्देहा बहवस्ते तु पश्चात्यानां प्रकीर्तिताः ॥ 132 ॥

naukāyanto jale yānaṁ jaghanyamiti gadyate |
taddehā bahavaste tu paśchātyānāṁ prakīrtitāḥ || 132 ||

Droṇī -Vessels having the shape of droṇī (pitcher).

Ghaṭī -Vessels built in the shape of ghaṭī (waterpot).

द्रोणीरूपन्तु यद्यानं द्रोणीयानं तदुच्यते ।
घटीभिर्घटितं यानं घटी नौकेति गद्यते ॥ 133 ॥

droṇīrūpantu yadyānaṁ droṇīyānaṁ taduchyate |
ghaṭībhirghaṭitaṁ yānaṁ ghaṭī nauketi gadyate ||

Phala-yāna- Vessels constructed with the kind of a gourd and the like were known as Phala-yāna.

Carma-yāna- Vessels made of thick leather were called Carma-yana (leather-vessels).

तुम्ब्याद्यैस्तुफलैर्यानं फलयानं प्रचक्षते ।
चर्मभिः स्थूलपूर्णैयच्चर्मयानं तदुच्यते ॥ 134 ॥

tumbyādyaistuphalairyānaṁ phalayānaṁ prachakṣate |
charmabhiḥ sthūlapūrṇaiyachcharmayānaṁ taduchyate ||

Vṛkṣa-yāna- Vessels constructed with light trees were known as Vṛkṣayāna (tree-vessels).

Jantu-yāna- Conveyance through water using animals was known as Jantu-yāna (animal vessels).

यानं यल्लघुभिर्वृक्षैर्वृक्षयानं तदुच्यते ।
जन्तुभिः सलिले यानं जन्तुयानं प्रचक्षते ॥ 135 ॥

yānaṁ yallaghubhivṛkṣairvṛkṣayānaṁ taduchyate |
jantubhiḥ salile yānaṁ jantuyānaṁ prachakṣate || 135 ||

The art of swimming does not fall within the category of low-quality vessels.

बाहुभ्यां सन्तरेद् वारि जघन्येषु तद्विर्णयः ॥ 136 ॥

bāhubhyāṁ santared vāri jaghanyeṣu tadnirṇayaḥ || 136 ||

In a nutshell, the elaborate description given in Yukti Kalpa Taru regarding the size, materials used, designs, technology, and other aspects of the ships proves beyond doubt that the tradition of marine engineering scaled remarkable heights in ancient India. It also proves beyond doubt that ancient Indians were seafaring people who colonised several countries of the world located on seashores. According to the literature, ships could carry crews numbering between 100 to 600. Out of regard for passenger convenience and comfort, the ships were well-furnished and decorated. Gold, silver, copper, and compounds of all these substances were generally used for ornamentation and decoration. The magnetic compass was first used in India for navigational purposes at sea and was known as 'Matsya yantra' (as pointed out above), because of the placement of a metallic piece shaped into a fish in a cup of oil.

Due to robust maritime technology, India could export its products worldwide in its ancient past. However, in the medieval period, a reversal plagued Indian society and the situation reached such a point that travelling across seas was looked down upon, and a person daring to do so was extricated from society and termed an outcast. On account of this social evil, India had to suffer a lot in commerce and shipping.

Archaeological evidence

From the archaeological horizon, the representation of ships on a seal and coin indicates maritime activities; there is enough evidence to show that the people of the Sindhu region carried on trade, not only with other parts of India but also with Sumer and the centres of culture in Western Asia, Egypt and Crete (Behera, 1999). At Lothal, a tidal dock believed to have been built during the Indus Valley Civilization (2300 BCE), near the present-day Mangrol harbour on the Gujarat (west) coast, is an example of early Indian seafaring. According to Rao (1979, 1985; 1987; 1991), the dock has been used in two stages: at the first stage, it was designed to allow ships 18-20 meters long and 4-6 meters wide, at least two ships could simultaneously pass and enter quickly; in the second stage, the inlet channel was narrowed to accommodate large ships but only single ships with flat bottoms could enter. The terracotta models of a boat from Lothal and engravings on Indus seals give some idea of ships going to the sea (Rao 1979, 1985). Further, the Indus Valley ports were set up to trade overseas with ancient Mesopotamia (the land between the Tigris and Euphrates rivers corresponding to the regions of modern Iraq, northeastern Syria, and southeastern Turkey) and Arabia along the Gulf of Cambay. These ports trans-shipped many sought-after Indian products from upriver cities along ancient Indus Valley rivers to the world market (Moorkerji, 1999). The distance from the mouth of the Sindhu River to Mesopotamia was approximately 2000 kilometres (km) and probably covered by Sindhu merchants from sites such as Dholavira and Mohenjo-Daro by sailing along the coast to various ports of the Arabian Gulf and Mesopotamia. From the above information, one can note concrete evidence of the seafaring and technology involved in constructing vessels in ancient India.

Further, there was a time when Indians were the masters of the seaborne trade in Europe, Asia, and Africa (Chanda, 1977). They built ships, navigated the sea, and held in their hands all the threads of international commerce, whether carried overland or by sea. Indian traders sailed their ships not only on the Indian Ocean and the Persian Gulf but also ventured into the Red Sea and even the Mediterranean and Aegean Sea (Vincent, 1998). From the very beginning, Indian traders had a very fair knowledge of all the ancient oceans and seas of the populated world (Radha, 1912). The Egyptians called India' God's land' because India was, in those days, significantly culturally developed (Prasad, 1990). The priests of ancient Egypt required vast quantities of aromatic plants for burning incense, and frankincense, myrrh, and lavender were also used for embalmment. Herodotus (around 425 BCE) has left us a sickening description of the many spices and scented ointments of which India was the centre (Marincola, 2003). Beauty products from India also attracted the women of Egypt, and the cosmetic trade depended entirely on imports chiefly from India (Prasad, 1977). The Pharaohs of the fifth and sixth dynasties made great efforts to develop trade relations with the land of Punt (Prasad, 1977). Knemphotep made voyages to Punt eleven times under the captainship of Koui (Mitra, 2007); these expeditions were organized and financed by the celebrated Queen Halshepsut. The vast extent of Indian cultural influences, from Central Asia to the north to tropical Indonesia, including the Philippines in the south and from the borderlands of Persia in the west to China and Japan in the east, has shown that ancient India was a radiating central civilization. Its advanced cultural thoughts, art, and literature were destined to leave deep marks on the races wholly diverse and scattered over the more significant part of Asia (as defined by the 'United Nations Geoscheme for Asia). The most valuable of the exports of India was silk, which under the Persian Empire is said to have been exchanged by its weight in gold. There was a huge consumption of Indian manufacturers in Rome (Warminton, 1974). Roman coins in large quantities are found in places in India (Gupta, 1965; Singh, 1988), whence beryl, pepper, pearls, and minerals were exported to Rome (Begley & De Puma, 1992). The chief articles of export from India were spices, perfumes, medicinal herbs, pigments, pearls, precious stones (like diamonds, sapphires, turquoise, and lapis lazuli), animal skins, cotton cloth, silk yarn, muslin, indigo, ivory, porcelain and tortoiseshell (De Silva, 1970). The chief imports were cloth, linen, perfume, medicinal herbs, glass vessels, silver, gold, copper, tin, lead, pigment, precious stones, and coral (Chanda, 1977).

Memoirs of Foreign Traveller

From the memoirs of foreign travellers, we get much information regarding the origin of the development of the shipping industry in India. For instance, Niccolo Di Conti (1395-1469), an Italian merchant, explorer, and writer, visited India during the early 15th century. He (Major, 1857: p. 27) writes that they build some ships much larger than ours, capable of containing 2,000 butts (2000 tons) and with five sails and as many masts. The lower part is constructed with triple planks to withstand the force of the tempest to which they are much exposed. However, some ships have built-in compartments that should one part be shattered, the other portion remaining intact to accomplish the voyage.

Venetian trader and explorer Marco Polo (1254-1324), had given a detailed description (John Masefield, 2003: Book 3, Chapter 1) of Indian ships, which is noteworthy. He observes the following:

"Having treated, in the preceding parts of our work, various provinces and regions, we shall now take leave of them and proceed to the account of India, the admirable circumstances of which shall be related. We shall commence with a description of the ships employed by the merchants, which are built of fir–timber.[65] They have

---

[65] The vegetable productions, and especially the timber, of

a single deck, and below this, the space is divided into about sixty small cabins, fewer or more, according to the size of the vessels, each of them affording accommodation for one merchant.[66] They are provided with a good helm. They have four masts with as many sails, and some have two masts that can be set up and lowered again as necessary.[67] Some ships of the larger class have, besides (the cabins), the number of thirteen bulk-heads or divisions in the hold, formed of thick planks let into each other. The object of these is to guard against accidents that may occasion the vessel to spring a leak, such as striking a rock or receiving a stroke from a whale, a circumstance that not infrequently occurs; for,

---

southern or maritime India, being different from the kinds known in Europe, it is improperly (if our author is actually speaking of Indian ships) that the ship–timber is said in the text to be the abete and zapino, as neither the abies nor pinus are found (in any accessible situation) between the tropies. But, irregular as it may seem, there will in the sequel be found reason to conclude that he is describing ships built in China, although for the Indian trade.

66   In the Latin of the Basle edition the number of these cabins is stated at forty, and they are said to be upon, not beneath, the upper deck. We know little of the interior of Indian vessels before the period of European intercourse, but in modern times their cabins are usually upon the after part of the quarter deck.

67   No mention is made of top masts in any modern description of Chinese junks; nor is it clear that such are here meant. The expressions may rather be understood of masts capable of being raised or lowered in the manner of those belonging to our lighters, and the sense of the passage may be- "They have four masts (with as many sails); two of which may be set up or lowered, as occasion may require."

when sailing at night, the motion through the waves causes a white foam that attracts the notice of the hungry animal. In expectation of meeting with food, it rushes violently to the spot, strikes the ship, and often forces in some part of the bottom. The water, running in at the place where the injury has been sustained, makes its way to the well, which is always kept clear. Upon discovering the situation of the leak, the crew immediately removed the goods from the division affected by the water, which, as a consequence of the boards being so well fitted, could not pass from one division to another. They then repair the damage and return the goods to that place in the hold from when they were taken. The ships are all double–planked; they have a course of sheathing boards laid over the planking in every part. These are caulked with oakum both inside and without and are fastened with iron nails. They are not coated with pitch, as the country does not produce that article, but the bottoms are smeared over with the following preparation. The people take quick–lime and hemp, which latter they cut small, and with these, when pounded together, they mix oil procured from a particular tree, making of the whole a kind of unguent, which retains its viscous properties more firmly and is a better material than pitch.[68]

---

[68]   This mode of preserving the bottoms of their vessels is common to the Chinese and the Indians. "At Surat," says Grose, "they excel in the art of ship-building. Their bottoms and sides are composed of planks let into one another, in the nature, as I apprehend, of what is called rabbet-work, so that the seams are impenetrable. They have also a peculiar way of preserving their ships' bottoms, by occasionally rubbing into them an oil they call wood-oil, which the planks imbibe." (Voyage to the East Indies, vol. 1. p. 107.) The mixture of chunam or lime with a

Ships of the largest size require a crew of three hundred men; others, two hundred; and some, one hundred and fifty only, according to their greater or less bulk. They carry from five to six thousand baskets (or mat bags) of pepper. In former times, they were of greater burthen than they are at present, but the violence of the sea having in many places broken up the islands, and especially in some of the principal ports, there is a want of depth of water for vessels of such draught, and they have on that account been built, in latter times, of a smaller size. The vessels are likewise moved with oars or sweeps, each requiring four men to work it. Those of the larger class are accompanied by two or three large barks, capable of containing about one thousand baskets of pepper, and are staffed by sixty, eighty, or one hundred sailors. These tiny crafts are often employed to tow the larger when working their oars, or even under sail, provided the wind be on the quarter, but not when right aft, because, in that case, the sails of the larger vessel must becalm those of the smaller, which would, in consequence, be run down. The ships also carry with them as many as ten small boats for anchors, fishing, and other services. They are slung over the sides and lowered into the water when there is an occasion to use them. The barks are, in like manner, provided with their small boats. When a ship, having been on a voyage for a year or more, needs repair, the practice is to give her a course

---

resinous oil, or with melted dammar, is commonly known in the dockyards of India by the name of gul–gul. "There would be no exaggeration" adds Grose, "in averring that they (the natives) build incomparably the best ships in the world for duration, and that of any size, even to a thousand tons and upwards it is not uncommon for one of them to last a century." -P. 108

of sheathing over the original boarding, forming a third course, which is caulked and paid in the same manner as the others; and this, when she needs further repairs, is repeated, even to the number of six layers, after which she is condemned as unserviceable and not sea–worthy.

Portuguese explorer Vasco da Gama, credited with discovering the sea route to India, followed a Gujarati trader from Zanzibar. Suresh Soni, the author of 'India's Scientific Heritage', quoting archaeologist Dr. Vishnu Shridhar Wakankar, said, "He no doubt came to India but not as a discoverer sea-farer but following a Gujarati trader from Zanzibar." According to Dr Wakankar, Vasco da Gama had recorded in his diary that he saw a docked ship three times bigger than his own upon his arrival at Zanzibar in Africa. He took an African interpreter to meet the ship's owner, Chandan, a Gujarati trader who used to bring pine wood and teak from India along with spices and take back diamonds to Cochin. Vasco da Gama followed Chandan to reach the shores of India, a fact very few in independent India know about, regrets Soni.

The author said Venetian trader and explorer Marco Polo, as early as the 13th century, had recounted that ships in India had double boards that were joined together with solid nails and crevices, filled with a special kind of gum and were so huge that 300 boatmen were needed to row them. These vessels could take a load of 3000 to 4000 gunny bags, having small rooms and arrangements for comfort. Additional layers were added to the bottom when it got damaged. Some ships had as many as six layers, the book says.

Another traveller, Berthma, had written how wooden

boards were joined to prevent even a drop of water from seeping into the ship and that it would take eight days to come to Iran from Cape Comorin (Kanyakumari), the book records.

9

Aeronautics

The science of aeronautics is the most crucial invention registered in the Vedic and post-Vedic literature. It can be said without hesitation that this science reached the hallmark of perfection. We have a lot of references from Vedas that speak of marine and aerial navigation. For instance, the following mantras are noteworthy.

तुग्रो ह भुज्युमश्विनोदमेघे रयिं न कश्चिन्ममृवाँ अवाहाः ।
तमूहथुर्नौभिरात्मन्वतीभिरन्तरिक्षप्रुद्भिरपोदकाभिः ॥3॥

*tugro ha bhujyumaśvinodameghe rayiṁ na
kaśchinmamṛvāṁ avāhāḥ ।
tamūhathurnaubhirātmanvatībhirantarikṣaprudbhirapodakā
bhiḥ ॥3*

(*Tugra*) A Vaiśya (merchant) desirous of riches (*rayim*) and prosperity (*bhujyum*) should go to far-off places for business (*asvina*) using surface, water, and air transport. He should (*avāhāḥ*) follow (*udmeghe*) sea routes (*ātmavantibhiḥ naubhiḥ*) by his own ships (*apodkābhiḥ*) which should be waterproof and rustproof, and aerial route by (*antrikṣa prudbhi*) aerial vehicle. (*kaśchit*) Anybody who does business in distant places, (*tam*) he (*ūhathuḥ*) gets the pleasure of visiting new and new places and (*na*) never (*mamṛvān*) fails in his enterprise.

The above mantra is talking about private ships and

aerial vehicles for merchants to visit far-off places in connection with the business.

तिस्रः क्षपस्त्रिरहातिव्रजद्भिर्नासत्या भुज्युमूहथुः पतङ्गैः ।
समुद्रस्य धन्वन्नार्द्रस्य पारे त्रिभी रथैः शतपद्भिः षळश्वैः ॥4 ॥

tisraḥ kṣapastrirahātivrajadbhirnāsatyā bhujyumūhathuḥ
pataṅgaiḥ |
samudrasya dhanvannārdrasya pāre tribhī rathaiḥ
śatapadbhiḥ ṣaḷaśvaiḥ ||4 ||

(*Tribhir-rathaiḥ*) Three kinds of transport- surface, marine, and air are required (pāre) for moving/travelling on earth (ārdrasya samudrasya) sea full of water and (dhanvanaḥ) space. These three means of transport should be (ati-vrajadbhiḥ) fast-moving as if (śatapadbhiḥ) propelled by 100 wheels (in case of surface transport), 100 paddles (in case of a ship) and 100 propellers (in case of airplane) (pataṅgaiḥ) fitted with one or more engines and (ṣaḍaśvaiḥ) and six powers of acceleration (gears). These vehicles may help you reach your destination within (tisraḥ) three (kṣapaḥ) nights or (triḥ) three (aha) days.

अनारम्भणे तदवीरयेथामनास्थाने अग्रभणे समुद्रे ।
यदश्विना ऊहथुर्भुज्युमस्तं शतारित्रां नावमातस्थिवांसम् ॥5 ॥

anārambhaṇe tadavīrayethāmanāsthāne agrabhaṇe
samudre |
yadaśvinā ūhathurbhujyumastaṁ śatāritrāṁ
nāvamātasthivāṁsam ||5 ||

(Aśvinau) O merchants and traders you (avīrayethā) venture into (samudre) in the ocean of water and in the space (tad) where there is (anārambhaṇe) no support, (anāsthāne) nothing to rest upon, (agrabhaṇe) nothing to cling to (ātasthivānsam) by boarding the (nāvaṁ) the ship (śatāritrā) (astam) steered by hundreds of propellers/paddles (ūhathuḥ) for carrying the (bhujyum) products or items of use from one country to another.

The Rāmāyaṇa, the Mahābhārata, the Purāṇas, and other scriptures abound in glowing descriptions of aerial travels.

Bhṛgu Saṁhitā classifies the vehicle flying in the sky as Agniyāna and the vehicle flying in space as Vyomyāna.[69]

Concept of invisibility of Vimānas described in Araṇya Kāṇḍa of Rāmāyaṇa: An aerial vehicle is described in Rāmāyaṇa for the first time in Āraṇya Kāṇḍa (3.49.19) when Rāvaṇa abducts Sitā.

स च मायामयो दिव्यः खरयुक्तः खरस्वनः ।
प्रत्यदृश्यत हेमांगो रावणस्य महारथः ॥

sa cha māyāmayo divyaḥ kharayuktaḥ kharasvanaḥ |
pratyadṛśyata hemāṁgo rāvaṇasya mahārathaḥ ||

It is described as 'The miraculous aircraft of Rāvaṇa, which is designed to appear and disappear at the wish of its master, yoked with miraculous mules, and built with its golden wheels and parts, appeared before Rāvaṇa braying noisily'.

Here, the critical aspect the author stressed is that 'the aircraft is designed in such a way that it could appear and disappear based on the wish of its master.' We need to think of the technology behind this feature. As per many ancient texts on Vaimānika Sāstra, vimānas in the Tretā

---

[69] जलेनौकेव यानं स्याद् भूमियानं रथं स्मृतम् ।
आकाशे अग्रियानं च व्योमयानं तदेव च ॥
jalenaukeva yānaṁ syād bhūmiyānaṁ rathaṁ smṛtam |
ākāśe agniyānaṁ cha vyomayānaṁ tadeva cha ||

Yuga were of the "Mantrika" (thought-powered) category. They used to be operated using mantra (thought power).

The science of aeronautics remained developed to its advanced stage until the *Mahābhārata* period. According to the description available in Droṇaparva, Vimānas was shaped like a sphere and could move along at great speed on a mighty wind generated by mercury. It says the vimāna moved like a UFO, going up, down, backwards and forwards as the pilot desired.

These texts describe two secrets to make the vimÈnas invisible. They are as follows.

Gūḍha: As explained in 'Vāyutatva Prakaraṇa' by Siddhanātha, quoted in by Bodhānanda Yati in his commentary on Bṛhadvimānaśāstra (Sūra, 1.2), by harnessing the powers, Yāsa, Viyāsaa, Prayāsa in the eighth atmospheric layer covering the earth, to attract the dark content of the solar ray, and use it to hide the vimÈna from the enemy.

गूढरहस्यो नाम- वायुतत्त्वप्रकरणोक्तरीत्या वातस्तम्भाष्टमपरिधि- रेखापथस्य यासावियासाप्रयासादि-वातशक्तिभि: सूर्यरिणान्तर्गततमश्शक्ति माकृप्य तत्संजोजनद्वारा विमानाच्छादन- रहस्यम् ॥ 5 ॥

gūṛharahasyo nāma- vāyutattvaprakaraṇoktarītyā vātastambhāṣṭamaparidhi- rekhāpathasya yāsāviyāsāprayāsādi-vātaśaktibhiḥ sūryariṇāntargatatamaśśakti mākṛpya tatsaṁjojanadvārā vimānāchchhādana- rahasyam ॥ 5

Adṛśya: As method laid down by 'Śakti-tantra', using the Vynarathya Vikaraṇa and other powers in the heart centre of the solar mass, attract the force of the ethereal flow in the sky, and mingle it with the balāhā vikaraṇa śakti in the aerial globe, producing thereby a white cover, which will make the vimana invisible (Commentary of Bodhānanda Yati on *Bṛhadvimānaśāstra, Sūra*, 1.2).

अदृश्यरहस्यो नाम - शक्तितन्त्रोक्तरीत्या सूर्यरथेषादण्ड-प्राङ्मुखपृष्ठकेन्द्रस्थवैणरथ्यविकरणादिशक्तिभिराकाशतरंगस्य शक्तिप्रवाह- माकृष्य वातमण्डलस्थबलाहाविकरणादिशक्तिपंचके नियोज्य तद्द्वारा श्वेताभ्रमण्डलाकारं कृत्वा तदावरणाद्विमानादृश्यकरणरहस्यम् ॥ 6 ॥

adṛśyarahasyo nāma - śaktitantroktarītyā sūryaratheṣādaṇḍa-prāṅmukhapṛṣṭhakendrasthavaiṇarathyavikaraṇādiśaktibhirāk āśataraṁgasya śaktipravāha- mākṛṣya vātamaṇḍalasthabalāhāvikaraṇādiśaktipaṁchake niyojya taddvārā śvetābhramaṇḍalākāraṁ kṛtvā tadāvaraṇādvimānādṛśyakaraṇarahasyam ॥ 6 ॥

A research article on the invisibility of aircraft (Hemant, 2015, pp. 37-44) states that by studying sunlight and types of sun rays, we can extract the power from the sun to make the aircraft invisible like the ancient vimānas. It also states that aircraft can be made invisible by the discovery of energy extraction from the dark matter in solar rays. Now, many military aircraft use new stealth technology, which makes aeroplanes invisible to radar and partially to the human eye, just like an invisibility cloak in Hollywood science fiction movies. They use nano-enabled coatings, which makes the aircraft invisible. Some Military aircraft are painted to match the sky from below and either match the ground or break up the aircraft's outline when viewed from above. Aircraft designers use camouflage and suppression techniques to reduce flight vulnerability to a missile attack. These include reducing the aircraft's radar reflectivity using non-reflecting materials and radar-absorbing paint when practical. (Shaw, 1985:55) The best example of military camouflage is the Canadian CF-18, as shown below:

Canadian CF-18 with False Canopy painted on the underside (all CF-18s are painted this way) to give the effect of the false cockpit to confuse the enemy is cited below.

The concept of flight of the Puṣpaka Vimāna based on 'Human Thought power' as described in Sundara Kāṇḍa: Puṣpaka vimāna is described as the best of the best aerial cars which are capable of travelling long distances. It is described as decorated with precious diamonds, gems, and corals with carvings of animals, birds, and flowers in gold and silver. It is said that Hanumāna, the son of Pavana, was surprised by looking into the tremendous aerial plane. Viśvakarmā, who has manufactured this, has praised this vimāna as the one without comparison in beauty. Every small part in this vimāna is made with great effort, with the best diamonds, and everything is of great significance.

तपः समाधान पराक्रमार्जितं मनः समाधान विचारचारिणम् ।
अनेक संस्थान विशेष निर्मितं ततस्तत्तुल्य विशेषदर्शनम् ॥ 5.8.4

tapaḥ samādhāna parākramārjitaṁ manaḥ samādhāna vichārachāriṇam /
aneka saṁsthāna viśeṣa nirmitaṁ tatastattulya viśeṣadarśanam // 5.8.4

Puṣpaka vimāna is a thought-powered vimāna; it is the one that moves about by thoughts of a concentrated mind, and it is made from various significant parts with an appearance of parts of equal significance, collected from here and there all over. The speciality of Puṣpaka vimāna is that 'it moves about by the thought power of its master.' Even though we do not have data regarding the exact technology behind this aircraft, recent research gives us an idea regarding probable technologies behind the construction of such aircraft. A few years ago, in 2013, researchers from the University of Minnesota designed a quadcopter that takes off using a noninvasive motor imagery-based brain-computer interface. This is the first time researchers have attempted to control a flying device through human thought. They have used a technique called electroencephalography to record the brain waves. These recordings are then transmitted to the quadcopter via WiFi to control it. (Lafleur 2013) Our ancestors might have used similar techniques to fly pushpaka vimana.

IV. Rāvaṇa's Vimāna similar to modern Jetpack: In the cave temples of Ellora, we can see the carvings of Rāvaṇa with a flying machine tied to his back. It represents a flying machine which Rāvaṇa had used in the era of

Rāmāyaṇa. Figure 3 below represents the same. It is comparable with our most advanced and fascinating flying devices of modern times called Jetpacks.

Fig 3

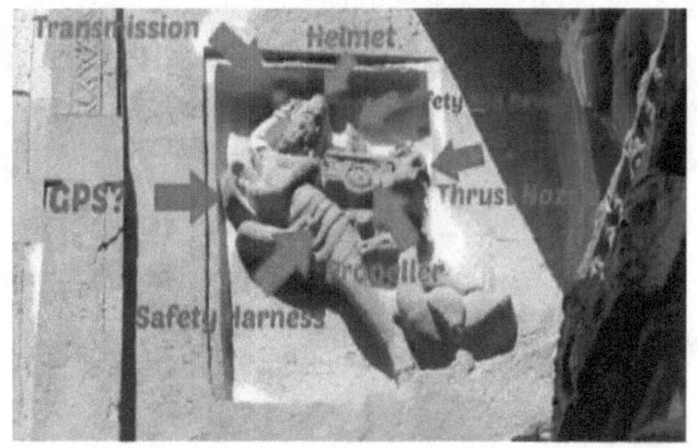

A 1200 Year Old Vimāna (Ellora Caves)[70]

Safety harness through which he has fastened the flying device to his back. Even in modern Jetpacks, we have six straps to tie the Jetpack to our back. The bird at the bottom represents the bird Jataayu, which fought with Rāvaṇa while he was travelling to Laṅkā after abducting Sīta. The circular wheel in the image represents a propeller similar to that used in modern Jetpacks. Modern Jetpack has two propellers; the image shown in Figure 3 represents a side view of the flying device. Assuming a circular wheel on the other side of the flying device, this is similar to our modern Jetpack. We can see a horse in the opposite direction of that of Rāvaṇa.

Looking at the size of the horse, in comparison with that of Rāvaṇa, it does not seem to be a real horse; it is

---

[70] Source: https://www.youtube.com/watch?v=rym865dwwri

comparable with the horse-shaped nozzle through which exhaust gases are thrown out of the flying device in the opposite direction of the flight which helps in an upward movement of the device. Even in modern Jetpacks, we see the two nozzle valves through which exhaust gases are thrown out of the device. The rectangular device above the wheels represents a safety roll device that is similar to the ones that are used in modern Jetpacks. In Figure 3, we can see Rāvaṇa wearing a crown and holding a device in his right hand near his ear. This is similar to the safety helmet pilots wear and the transmission device through which pilots interact with air stations to receive the signals for takeoff or landing. It is described that Rāvaṇa had multiple flying machines, which he regularly used for travelling to various places and had six airports in Laṅkā where he used to park these aircraft. Rāvaṇa might have been interacting with these airport base stations using the transmission device he holds in his right hand. The device he is using in his left hand is comparable with the GPS device the pilots use even today. This makes us think if such a flying device with technology similar to that of our modern times existed during the era of Ramayana.

Mantra-powered aircraft simulation and speed of Puṣpaka Vimāna as described in Yuddha Kāṇḍa

We can find the description of Puṣpaka Vimāna once again in Yuddha Kāṇḍa when Rāma kills Rāvaṇa and decides to return to Ayodhya. Rāma asks Vibhiṣaṇa to see how they can quickly reach Ayodhyā as travelling back on the same path would be very difficult. Then Vibhiṣaṇa suggests that he can drop all of them to the city within a day and describes Puṣpaka Vimāna to him.

एवमुक्तस्तु काकुतस्तं प्रत्युवाच विभीषणः ।
अह्ना त्वां प्रापयिष्यामि तां पुरीं पार्थिवात्मज ॥ 6.121.8

ēvamuktastu kākutastaṁ pratyuvācha vibhīṣaṇaḥ /
ahnā tvāṁ prāpayiṣyāmi tāṁ purīṁ pārthivātmaja // 6.121.8

Hearing the words of Rāma, Vibhiṣaṇa replied: "O prince! I will get you to that City in a day."

The main point that we need to consider here is that the Puṣpaka Vimāna described in the Rāmāyaṇa could cover a distance of Laṅka to Ayodhyā within a day. The aerial distance between Lanka and Ayodhya is around 2110 km, i.e. 1311 miles. That means Pṣhpaka Vimāna in those days could cover this aerial distance of 2110 km within a day. There is no description of Vimāna travelling during the night in Rāmāyaṇa hence, we can probably guess that Puṣpaka vimāna might have taken a maximum of 6 to 8 hours during day time to travel from Laṅkā to Ayodhya. Both Rāma and Lakṣmaṇa were amazed by seeing this vimāna. Śloka below describes the same..

तत्पुष्पकं कामगमं विमानमुपस्थितं भूधर सन्निकाशम् ।
दृष्ट्वा तदा विस्मयमाजगाम रामः सौमित्रिरुदारसत्त्वः ॥ 6.121.30

tatpuṣpakaṁ kāmagamaṁ vimānamupasthitaṁ bhūdhara
sannikāśam /
dṛṣṭvā tadā vismayamājagāma rāmaḥ
saumitrirudārasattvaḥ // 6.121.30

The generous-minded Rāma, along with Lakṣmaṇa, felt   amazed to see the aforementioned aerial car, Pushpaka, which resembled a mountain and could travel everywhere at will, arrived on that occasion.

The critical point for us here is 'kāma gamam vimānam', i.e. this puṣpaka vimāna could 'travel everywhere at will'. Here, once again, we can compare this with thought-powered aircraft. After the research by Prof. Bin's team from the University of Minnesota in the year 2013, Tim Fricke, along with his team of German scientists from the Technical University of Munich, researched a flight simulator in May 2014 using which aviators could use nothing but their brains to fly the aircraft. In this project, just using their thought power, a team of seven pilots, some with no previous flying experience, could control the flight simulator. Electrical signals from their brain were being read by their helmet, which had dozens of electroencephalography electrodes. They used an algorithm that would convert the brain signals into computer commands. The pilots who underwent this experiment could control the plane well enough to satisfy all the criteria for gaining a pilot license. Tim Fricke, an aerospace engineer who heads the project at the Technical University of Munich, said his long-term vision is to make flying accessible to more people. (Ellie 2014)

When we analyze the descriptions of vimanas in various kāṇḍas of Rāmāyaṇa, four critical technical aspects can be compared with modern aircraft. They are:

a). The invisibility of aircraft. It is described that Rāvaṇa's vimāna was designed to appear and disappear at the wish of its master. Maharshi Bharadwaja's Bṛhad Vimāna Śāstra also describes the concept of 'Gūḍha' and 'Adṛṣya', which can be compared with the concept of invisibility. Even our modern military aircraft use stealth technology and camouflaging techniques to achieve the same.

b). Flight based on the thought power of the pilot.

This is a fascinating concept that was like fantasy till recent years. However, there is extensive research going on on this subject across the world. Research done by the University of Minnesota and the Technical University of Munich in recent years makes us think of probable technology behind this concept in vimānas described in Rāmāyaṇa.

c). Comparison between Rāvaṇa's vimāna with modern Jetpack. Carvings of Rāvana's vimāna in Ellora caves are in many ways comparable to modern Jetpacks. This makes us think of the advancements of technology in that era. Even though we do not know the exact technology behind Rāvaṇa's Jetpack, there is definitely some

d). Seating capacity of Puṣpaka vimāna. Another critical aspect of Puṣpaka vimāna was its size; it could spaciously accommodate Rāma, Sītā, Lakṣmaṇa, Vibhiṣaṇa, Sugrīva, Vānara chiefs and their wives, Ṛkṣas and other Rākṣas. (Ravi Prakash Arya, 1970) The count of which easily might be around 200 to 400. This aspect of Pu–paka vimāna can be compared with our Airbus 380-800, which can accommodate 853 passengers. This is our modern-day Puṣpaka vimāna. By analysis of descriptions of vimānas in Rāmāyaṇa, we understand the fact that advanced technology prevailed during the era of Rāmāyaṇa. Even though we do not know the exact technology behind ancient vimānas, we can compare them with today's prevalent technology. However, the question that is even now unanswered is that we really do not know how ancient technology died out as the eras passed by.

Section 43 of *Vanaparva* describes Arjuna's arrival at

Indra's Amarāvati'. Here, not only vimÈnas are mentioned, but even an airport, where vimÈnas were stationed correctly and used to land and take off similarly as we see today in busy airports. The description is quoted below:

"And having beheld those celestial gardens resounding with celestial music, the strong-armed son of Pāṇḍu entered the favourite city of Indra. In addition, he beheld their celestial vimānas by thousands, capable of going everywhere at will, stationed in proper places. And he saw tens of thousands of such cars moving in every direction."

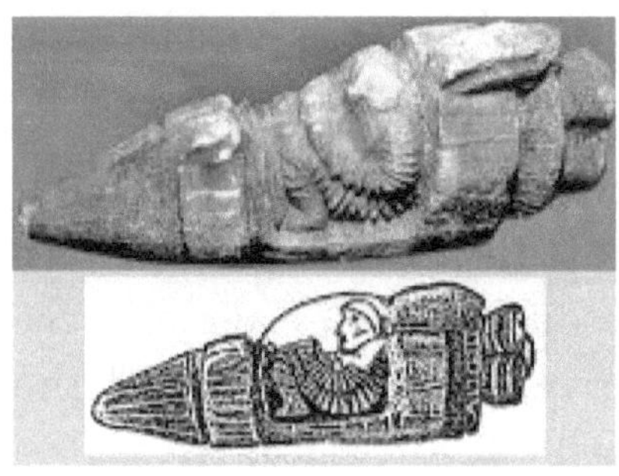

Also, section 165 of *Vanaparva* mentions aircraft as follows:

"Moreover, from all sides in vimānas resplendent as the sun, hosts of Gandharvas and Apsaras began to follow that repressor of foes, the lord of the celestials. In addition, ascending a vimāna yoked with engines powerful like a horse, decorated with burnished gold, and roaring like clouds, that king of the celestials, Purandara, blazing in beauty came unto the Pārtha. And having arrived (at that place), he was guarded by a thousand eyes of security men descended from his vimāna".

See also the description (section 164) when Arjuna was dropped by Śiva's spaceship piloted by "Mātali ".

"Furthermore, it came to pass that once days as those mighty charioteers were thinking of Arjuna, seeing Mahendra's vimāna, yoked with engines with the effulgence of lightning, arrive all of a sudden, they were delighted. Moreover, driven by Mātali, that blazing vimāna, suddenly illuminating the sky, looked like smokeless flaming tongues of fire, or a mighty meteor embosomed in clouds and seeing before them that vimāna driving in which the slayer of Namucī had annihilated seven phalanxes of Diti's offspring, the magnanimous Pārtha went round it. In addition, being highly pleased, they offered excellent worship unto Mātali, as unto the lord of the celestials himself. Moreover, the son of the Kuru king duly enquired of him after the health of all the gods, and Mātali also greeted them. And having instructed the Pārtha even as a father doth his sons, he ascended that incomparable vimāna and returned to the lord of the celestials".

The mention of "smokeless flaming tongues of fire" could be clearly perceived as an eco-friendly jet engine's emission.

The *Samarāṅgaṇa Sūtradhāra* [SS.] (Chapter, 31) of Rājā Bhoja (336-392 A.D.) sheds ample good light on the types and construction of the airplane. From the mention of an aerial machine in the other texts, only the barest details of its make-up could be gleaned, but the only text that gives us some knowledge of its actual construction is this work of Bhoja. Firstly, Bhoja (SS.31.95) explains that an airplane was made with a lighter body with a solid and fine frame. An external combustion engine was introduced in its centre with a combustion chamber below it.

लघुदारुमयं महाविहंगमं दृढसुश्लिष्टतनुं विधाय तस्य ।
उदरे रसयन्त्रमादधीत ज्वलनाधारमधोऽस्य चातिपूर्णम् ॥ 31.95

laghudārumayaṁ mahāvihaṁgamaṁ dṛḍhasuśliṣṭatanuṁ vidhāya tasya /
udare rasayantramādadhīta jvalanādhāramadho'sya chātipūrṇam // 31.95

Having created a vast bird-like lighter body with a solid and fine frame, introduce an external combustion engine in its centre with the combustion chamber below it.

The motive force is then explained:

तत्रारूढः पुरुषस्तस्य पक्षद्वन्द्वोच्चालप्रोज्झितेनानिलेन ।
सुप्तस्वान्तः पारदस्यास्य शक्त्या चित्रं कुर्वन्नम्बरे याति दूरम् ॥ 31.96

tatrārūḍhaḥ puruṣastasya pakṣadvandvochchālaprojjhitenānilena /
suptasvāntaḥ pāradasyāsya śaktyā chitraṁ kurvannambare yāti dūram // 31.96

Having boarded the airplane, the pilot should operate the two fans (mounted on a shaft) due to the power of mercury. The spinning of fans (wings) will create lift by the air released by them downward. It makes the airplane fly in a marvellous way far in the sky.

A big-sized *vimāna* is then described (SS. 31.97); it contains not one, as in the previous case, but four combustion chambers.

इत्थमेव सुरमन्दिरतुल्यं संचलत्यलघु दारुविमानम् ।
आदधीत विधिना चतुरोऽन्तस्य पारदभृतान् दृढकुम्भान् ॥ 31.97

itthameva suramandiratulyaṁ saṁchalatyalaghu dāruvimānam /
ādadhīta vidhinā chaturo'ntasya pāradabhṛtān dṛḍhakumbhān // 31.97

The big-sized Vimāna looks like the aerial vehicle of Devas when flying in the sky. It has four combustion chambers. A terrific noise was produced with the help of the engine of the airplane to scare away elephants on the battlefield and throw them completely out of control (SS. 31.98-100).

अयःकपालाहितमन्दवह्निप्रतप्तकुम्भभुवा गुणेन ।
व्योम्नो झगित्याभरणत्वमेति सन्तप्तगर्जद्रेरसराजशक्त्या ॥31.98

ayaḥkapālāhitamandavahniprataptakumbhabhuvā guṇena /
vyomno jhagityābharaṇatvameti santaptagarjadrarasarājaśktyā //31.98

वृत्तसन्धितमथाय सयन्त्रं तद् विधाय रसपूरितमन्तः ।
उच्चदेशविनिधापितप्तं सिंहनादमुरजं विदधाति ॥ 31.99

vṛttasandhitamathāya sayantraṁ tad vidhāya rasapūritamantaḥ /
uchchadeśavinidhāpitaptaṁ siṁhanādamurajaṁ vidadhāti //

स कोऽप्यस्य स्फारः स्फुरति नरसिंहस्य महिमा ।
पुरस्ताद् यस्यैता मदजलमुचोऽपि द्विपघटाः ॥

मुहुः श्रुत्वा श्रुत्वा निनदमपि गम्भीरविषमं ।
पलायन्ते भीतास्वरितमवधूयांकुशमपि ॥ 31.100

sa ko'pyasya sphārah sphurati narasimhasya mahimā /
purastād yasyaitā madajalamucho'pi dvipaghaṭāḥ //
muhuḥ śrutvā śrutvā ninadamapi gambhīraviṣamaṁ /
palāyante bhītāsvaritamavadhūyāṁkuśamapi // 31.100

This specific military use of aircraft against elephants tempts one to suggest that the *Hasti-yantra* advocated by Kauṭilya against elephants was something like the heavier vimāna described by Bhoja (336-392 A.D.).

Bhoja (SS. 79-81.) also mentions that technical expertise has not been revealed here for the sake of protecting the copyright of those expert families in whose tradition this technology was developed.

दुष्करं यद्यदन्यच्च तत्तद् यन्त्रात् प्रसिध्यति ।
यन्त्राणां घटना नोक्ता गुप्तर्थं नाज्ञतावशात् ॥79 ॥

duṣkaraṁ yadyadanyachcha tattad yantrāt prasidhyati /
yantrāṇāṁ ghaṭanā noktā guptarthaṁ nājñātāvaśāt //79 //

तत्र हेतुरयं ज्ञेयो व्यक्ता नैते फलप्रदाः ।
कथितान्यत्र बीजानि यन्त्राणां घटना न यत् ॥80 ॥

tanna heturayaṁ jñeyo vyaktā naite phalapradāḥ /
kathitānyatra bījāni yantrāṇāṁ ghaṭanā na yat //80 //

तस्माद् व्यक्तीकृतेष्वेषु न स्यात् स्वार्थो न कौतुकम् ।
वस्तुतः कथितं सर्वं बीजानामिह कीर्तनात् ॥81 ॥

tasmād vyaktīkṛteṣveṣu na syāt svārtho na kautukam /
vastutaḥ kathitaṁ sarvaṁ bījānāmiha kīrtanāt //81 //

This idea is also noticed in the *Bṛhatkathā* story.

An important point to be noted in Bhoja's treatment is that he discusses the views of some earlier writers

(31.87).

पारम्पर्यं कौशलं सौपदेशं
शास्त्राभ्यासो वास्तुकर्मोद्यमो धीः ।
सामग्रीयं निर्मला यस्य सोऽस्मिं-
श्चित्राण्येवं वेत्ति यन्त्राणि कर्तुम् ॥87॥

pāramparyaṁ kauśalaṁ saupadeśaṁ
śāstrābhyāso vāstukarmodyamo dhīḥ I
sāmagrīyaṁ nirmalā yasya so'simma-
śchitrāṇyevaṁ vetti yantrāṇi kartum II87 II

He expressly mentions some of his yantras as having been described by the ancients (31.84)

अग्रतश्च पुनर्ब्रूमः कथितं यत् पुरातनैः ।
बीजं चतुर्विधमिह प्रवदन्ति यन्त्रे-
ष्वम्भोग्निभूमिपवनैर्निहितैर्यथावत् ।
प्रत्येकतो बहुविधं हि विभागतः स्या-
न्मिश्रैर्गुणैः पुनरिदं गणनामपास्येत् ॥84॥

agrataścha punarbrūmaḥ kathitaṁ yat purātanaiḥ I
bījaṁ chaturvidhamiha pravadanti yantre-
ṣvambhognibhūmipavanairnihitairyathāvat I
pratyekato bahuvidhaṁ hi vibhāgataḥ syā-
nmiśrairguṇaiḥ punaridaṁ gaṇanāmapāsyet II84 II

He also refers to *yantraśāstra-adhikāra* as comprising five sections (SS. 31.88). All this implies the existence of a technical literature on the yantras. Dandin mentioned earlier authors on the subject, Brahma, Indra, and Parasara.

चित्रैर्युक्तं ये गुणः पंचरूपं
ज्ञानन्त्येनं यन्त्रशास्त्राधिकारम् ।
ये वा कृत्स्नं योजयन्तेऽत्र
सम्यक् तेषां कीर्त्तिर्घां भुवं चावृणोति ॥88॥

chitrairyuktaṁ ye guṇaḥ paṁcharūpaṁ

jñānantyenaṁ yantraśāstradhikāram I
ye vā kṛtsnaṁ yojayante'tra
samyak teṣāṁ kīrttirdyāṁ bhuvaṁ chāvṛṇoti II88 II

A further point of interest is that Bhoja (SS. 31.82) speaks of some of these as having been seen and described on first-hand knowledge of them.

अभ्यूहां स्वधिया प्राज्ञैर्यन्त्राणां कर्म यद् यथा।
यन्त्राणि यानि दृष्टानि कीर्तितान्यत्र तान्यपि ॥82 ॥

abhyūhyaṁ svadhiyā prājñairyantrāṇāṁ karma yad yathā I
yantrāṇi yāni dṛṣṭāni kīrtitānyatra tānyapi II82

The references to yantras found in highly reputed works cannot be doubted, and facts mentioned by one writer receive confirmation from another.

Apart from Bhoja (336-392 A.D.), Budhasvamin's version of *Brihatkatha* has essential information on aerial vehicles, *Ākāśa-yantras,* as he expressly calls them. The context where the description of the *Ākāśa-yantra* occurs is the longing of the pregnant Vāsavadattā; in the Kashmirian version of Somadeva (11th century A.D.), there is only a line saying that her *dohada* (longing) was fulfilled by manifold yantras, etc., arranged by the Minister Yaugandharāyaṇa. However, in Budhasvamin's version, the context contains an elaborate description of the yantras. Vāsavadattā yearned to see the whole world from above in the aerial vehicle (Śloka, 190); Rumanvān, the commander-in-chief, at once ordered carpenters to manufacture a flying yantra (Śloka, 196). The carpenters say that they know only four kinds of yantras, made respectively with water, stone, mud, and twigs; that it is the Yavanas (Indian kings of Northwestern region) who know the *Ākāśa-yantras* and that they, for their part, have not even laid eyes on them.

आकाशयन्त्राणि यवनाः किल जानते ।
अस्माकं तु यातानि न गोचरं चक्षुषामपि ॥

ākāśayantrāṇi yavanāḥ kila jānate /
asmākaṁ tu yātāni na gocharaṁ chakṣuṣāmapi //

Thereupon, a Brahman told Rumanvan a story to illustrate how, in the matter of the aerial vehicle, architects made a secret of their lore and uttered the falsehood that they knew it not.

With Mahasena, King of Ujjain and father of Vāsavadattā, was an architect named Pukkasaka who once went out to Saurāṣṭra along with the king's camp. He came across a young architect, Viśvila, who was considered equal to Viśvakarman in knowledge. To Viśvila's father, called Maya, Pukkasaka proposed that he desired to give his daughter Ratnāvali in marriage to his son. The proposal was agreed to, and Pukkasaka was waiting for his son-in-law's arrival. Once, after attending to his work, Pukkasaka returned relatively late and, to his surprise, found none in his house eager to attend to his bath and dinner.

On enquiry, he heard from his wife that a visitor had upset their home; the visitor had come over with some rice and asked that it be cooked for him. The house had been burning fuel, yet the rice would not moisten. Pukkasaka now understood that his son-in-law had arrived and desired to see the youth. Viśvila issued out of the workshop, and when the puzzled Pukkasaka asked him what that so-called rice was, Viśvila revealed that they were fake, delicate rice-like chiselling from the white wood of the Karaghāṭa tree. The marriage of

Viśvila and Ratnāvali was then celebrated. After a time, Viśvila learnt from his brothers-in-law their anxiety about feeding in their house one more family member, the son-in-law. Viśvila at once repaired to the forest cut down certain kinds of wood and manufactured out of them yāvana-machines यन्त्राणि घट्टयामास अथ विश्विलः (Śloka, 224) as also manifold household utensils conducive to health and longevity, according to the principles laid down in *Vrikṣāyurveda* (225); he sold these for thousands of Rupees and earned monetary benefit to his father-in-law.

Once, Pukkasaka sadly told his family of his impending departure for Benaras what place King Mahasena had ordered him to go to build a temple for his friend King Brahmadatta of Benaras. Viśvila requested the king to depute him in place of his father-in-law, and, with the king's permission and accompanied by a retinue, he departed for Benaras. At the end of every day's journey, however, Viśvila would vanish somewhere and slip back invisibly into the camp. At his home, RatnÈvali shortly became pregnant, to the surprise and agony of her parents. That news reached the king who, putting two and two together, explained that every night, Viśvila, according to the report of the men of his retinue, would mount a machine-cock, *Yantra-kukkuṭa,* fly away somewhere and stealthily and with shrouded face slip back into his bed in the small hours of the morning. Once he had been forced to return late in the morning and, confessing to his friends about his nocturnal visits to his wife by an aerial vehicle, he had begged them not to inform any architects or non-specialists of the secret of his aerial vehicle, as if that knowledge became public, the Akāśa-yantra would become a cheap affair like a cot.

आकाशयन्त्रविज्ञानं दुर्विज्ञानमयावनैः
खट्वाघटनज्ञानमिवेदं प्रचुरीभवेत् ।
लोकेन परिभूयते क्षणरागाः हि मानुषाः ॥ Ślokas 250-251)

ākāśayantravijñānaṁ durvijñānamayāvanaiḥ

khaṭvāghaṭanajñiānamivedaṁ prachurībhavet I

lokena paribhūyate kṣaṇarāgāḥ hi mānuṣāḥ II

That explained how Ratnāvali came to bear a child. Soon, the temple at Benaras was finished, and Viśvila returned.

The King would not believe his words and pressed him; after that, Pukkasaka pressed Viśvila. Viśvila pretended to reveal the secret, but that night, he woke up his wife and gave her an ultimatum; by pressing him for knowledge of the flying machine, her father was virtually driving him to his own home, that she had to choose between father and husband, and that, so far as he was concerned, he would give her up rather than the secret. She took little time to decide, and in a moment, they were off on the machine-cock:

यानं कुक्कुटसंस्थानमास्थाय सह भार्यया ।
रात्रावाकाशमुत्पत्य स्वस्थानं ययौ विश्विलौ ॥ (Ślokas 252)

yānaṁ kukkuṭasaṁsthānamāsthāya saha bhāryayā I

rātrāvākāśamutpatya svasthānaṁ yayau viśvilau II

Having told the story, the Brāhmaṇa told the commander-in-chief Rumaṇvān that śilpis (technocrats) keep their knowledge secret, that all of them might well be bound and beaten till they agreed to make the aerial vehicle. As Rumaṇvan was putting that advice into action, a new śilpi (techno cat) offered to manufacture an airplane.

The new *śilpin* asked Rumaṇvan to collect the

materials. When these were assembled and the work was to start, the old *śilpis* suggested to the new ones that they ascertain from Rumaṇvān the required seating capacity.

The mention of this deserves to be noted, as also the further observation that there had been cases in the past in which airplanes had been made without regard to seating capacity, and they had come to ruin, with the result that the Kings had cruelly dealt with their makers.

आरोहकपरीणाहं सेनानीरनुयुज्यताम् ।
अज्ञातवाह्यसंख्याभिः बहवः शिल्पिनो नृपैः ।
विपन्नयन्त्रैः श्रूयन्ते मथिताः कुपितैरिति ॥

ārohakaparīṇāhaṁ senānīranuyujyatām /
ajñātavāhyasaṁkhyābhiḥ bahavaḥ śilpino nṛpaiḥ /
vipannayantraiḥ śrūyante mathitāḥ kupitairiti //

The new *śilpi* replied that the airplane he was going to make was of a superior type quite different from the productions of the stupid silpis to whom they referred and that the seating capacity of his vehicle was not limited.

He made an airplane in the shape of Garuda accordingly. Vāsavadattā and Udayana mounted it with their retinue, roamed about, called on Padmavati's brother in Magadha on the east and Vāsavadattā's parents at Ujjain on the west and returned to Kausambi.

This fulfilment of the Queen's *dohada* (longing of a pregnant woman) during her pregnancy for an aerial flight also echoes in Jain *Kāvya* literature. In Vādbhāsiṅha's *Gadya-cintamani* and *Kshatra-chuḍāmani*, silpis make a peacock-like aerial car (*Mayura-yantra*) for the pleasure flight of Queen Vijayā. There is a story in the *Ratnaprabhālambaka,* 7. 9. It is about Karpurika in the city of Karpurasambhava, of whom NaravÈhanadatta has a dream. The Prince starts in search of her in the company of Gomukha. *En route,* they come to a Hemapura, where, as Naravāhanadatta is going along the bazaar street, he comes across everything about a city: shops, things, servants, men, and women, but their speechless movements and activities reveal to him the wondrous fact of their all being robots (10, 11). He then makes his way to the palace, where he finds the only sentient being sitting on a throne-like a king, and, like the soul presiding over the body and senses, manipulating the mechanical city and being served on all sides by his robotic servants. On being asked by Narav āhanadatta, the mystery man of this machine city recounts his story: He, Rājyadhara by name, and his elder brother Prāṇadhara, were originally residents of Kāñchlpura where King Bahubala (Mahabala in Kshemendra) was ruling.

Both were śilpis (engineers) and adepts in manufacturing electronic (magic) yantras, devised originally by Maya. It may be noted in passing that, according to Dandin, Kañchi had some śilpis adept at making yantras. The elder brother sought the company of courtesans and squandered his and his younger brother's property. Reaching the end of his material resources, the elder brother thought of theft and harnessed his skill in yantras for that purpose. He devised a pair of wooden swans to move along an electronic rope operated from one end. The other end of the rope was tied to the window of the King's treasury into which, night after night, these swans were sent; their beaks, which were put into action, removed the lid of jewel boxes and picked up some jewels and then the swans, were again moved back to their original place. This mysterious theft was going on, and the King ordered an all-night vigil to catch the culprits. The swans were seen doing this dexterous job and were detached from the rope, which suddenly sagged; the fastening nail became loose, and Prāṇadhara immediately understood that the theft had been discovered. He asked his brother to accompany him to a far-off place to escape being caught as thieves early in the morning.

Prāṇadhara said that he had an aerial vehicle that could, in a single sweep, fly across 800 *yojanas* and got into it immediately with his family. As the seating capacity of that yantra had been reached by the crowd that entered it, Rajyādhara left in his vehicle, a *vāta-yantra vimāna,* which he had made; that vimāna took him at one jerk over 200 *yojanas,* and with a second propelling over another 200 *yojanas.*

That brought the younger brother, the narrator, to the city of Hemapura. When he reached it, it was an abandoned place; he thought of peopling it with the help of his mechanical skill and created a yantra-population. The next day, he learnt from Naravāhanadatta the search that the latter was on and helped him with an aircraft that took him to the city of Karpūrasambhava. There, the Prince found the heroine of his dream, Karpurikā, and married her. When he desired to return home with his new bride, his father-in-law revealed that he, too, had a visitor-śilpi in his city who could provide him with an aircraft. It happened that the visitor-śilpi was no other than Prāṇadhara, the elder brother of Kañchi. The vehicle that Prāṇadhara gave him was a veritable flying fortress, as it could lightly bear a thousand passengers (228), and in that they returned home, touching Rājyadhara's city *en route.*

In Kshemendra's brief narration of this story (14. 459-508), the thieving swans are mentioned as many, the elder's aircraft is called a *Yantracakra,* and the younger's is said to possess double the speed of the elder's.

There are also two other contexts in the *Kathāsaritsāgara* where the aerial yantra figures (the story of Puṣkarākṣa and Vinayavatī and "the story of' Somaprabhā and the three suitors.

Further, the consistent reference to the Yavanas (Indians inhabiting Northwestern regions) as handling the Ākāśa-yantras not only adds realism to these descriptions.

Thus, the details mentioned in our works concerning the *Ākāśa-yantras,* such as seating capacity, travelling range, the difference in speed, accidents, materials and methods of manufacture, and various Kings' curiosity about and fondness for these, show that we are no longer in the world of imagery but in the world of fact.

The sage Bharadvāja wrote an outstanding work of far-reaching importance, named 'Yantra Sarvasva' consisting of as many as 40 sections, of which *Bṛhad Vimāna Śāstra* constitutes the most glaring work on aeronautics. Here it may be noted that Maharishi Dayananda, a great embodiment of Vedic life and thought and prominent social reformer of 19th century India, also talks about it in his fifth speech on the subject of the Veda delivered in Poona in 1875 to have come across a book on Vimāna Śāstra (Upadesh Manjari, 1998:

p.28):

" मैंने भी एक विमान रचना का पुस्तक देखा है । "

" maiṁne bhī ēka vimāna rachanā kā pustaka dekhā hai । "

I saw a book dealing with the manufacturing of "Aerial vehicles".

Hence, Maharshi Dayananda had a firm and positive knowledge of this matter. The extant *Bṛhad Vimāna Śāstra* consists of 8 chapters, a hundred topics and 500 sūtras. There is a luminous commentary on this treatise by Śri Bodhananda Yati, known as Bodhānanda vṛtti. The commentary consists of about 300 verses.

It is pretty evident from this book itself that long before this great and glorious treatise, sages like Nārāyaṇa, Śaunaka, Garga, Vāchaspati, Chakrāyaṇī, and Dhuṇḍinātha respectively produced brilliant scientific work in this field, namely Vimānachandrikā, Vyomayāna Yantra, Yantra Kalpa, Yānabindu, Kheṭayāna Pradipikā, Vyomayānārka Prakaśa.

From the exact text, we are given to understand that in addition to the above, the following sages had also worked in the field of aeronautics:

1. Viśvanātha   2. Gautama   3. Lalla

4. Viśvambhara   5. Agstya   6. Buḍila

7. Gobhila   8. Śākaṭāyana   9. Atri

10. Kapardī   11. Gālava   12. Agnimitra

13. Vātāpa   14. Sèmba   15. Bodhānanda

16. Bhardwāja   17. Sidhnātha   18. Īśvara

19. Āśvalāyana   20. Vyāsa   21. Parāśara

22. Siṅhakoṭa   23. Aṅgirātab   24. Visaraṇa

25. Vasiṣṭha   26. Jaimini   27.Āpastamba

28. Baudhāyana   29. Nārada   30. Vālmīki

An aeroplane is called 'Vimāna' because it behaves like a vi 'bird' in the sky.

Bṛhad Vimāna Śāstra defines the Vimāna as under under:

वेगसम्याद् विमानोण्डजानामिति । 1.1

vegasamyād vimānoṇḍajānāmiti I 1.1

"Due to the similarity of the principle in flying applied to the movement of birds, it is named Vimāna."

It can travel to distant aerial regions. Bharadvāja describes three classes of the airplanes:

1. Travelling from one place to another within the same country.

2. Travelling from one country to another country.

3. Travelling from one planet to another planet.

From this, it is pretty clear that the science of aeronautics was in a very advanced stage in ancient India, and space travel was very well known to Indians and that the spacecraft highly flourished in India. This also authenticates the stories regarding space travel and the concept of elongation of life in space travel, like the one described in the *Mahābhārata*. During that Vedic period, it was accepted as a proven fact that space travel with tremendous speed may cause the elongation of life. The story of Kakudmi and Revati, Balaram's wife, portrays this fact. Revati and her father, Kakudami, returned from space after 26 years. Revati was youthful, in puberty, and Kakudami lived to give her in marriage to Balaram. Modern science has not yet proved this fact, though Einstein has supported this concept theoretically in the 51st century of the Kali era.

There is also a detailed description of 25 types of Vimānas named after their form, capacity, and nature of the operation. Some may be mentioned as 'Tripura Vimāna', which could move quickly on land, in the sea, and the air. There were another three sorts of Vimānas called 'Śakuna', 'Rukma', and 'Sundara'.

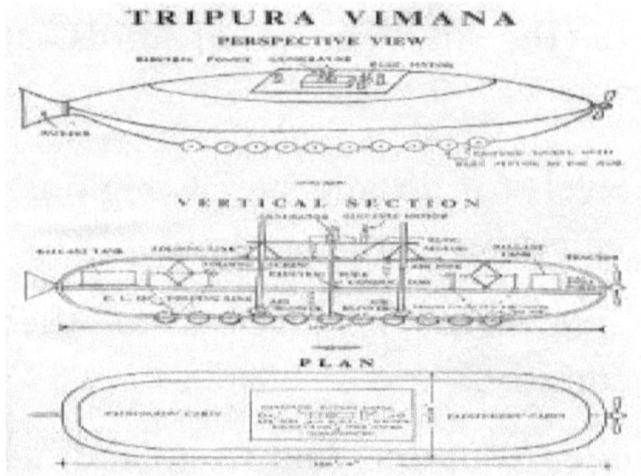

The experiments carried out based on the descriptions of *Bṛhad Vimāna Śāstra* were found extremely stunning and glaring to the scientists. CSR Prabhu of Hyderabad did some work on the technological devices mentioned in *Vaimānika Śāstra* and could recognise some of the 31 machines described in it successfully prepared the alloys as described in its chapter on metals.

Vimānas again are divided into three classes:

Māntrika- Operating at the click of an idea of a Yogi who has attained perfection in samprajñāta samādhi.

Tāntrika-Operated by divine powers of a seeker obtained by him during the observance of asamprajñāta samādhi.

Yāntrika-Machine operated

Let alone the first two classes of planes, the last class was highly advanced. The Yāntrika plane would have as many as 32 contrivances without the practical knowledge of the machinery of which aerial navigation would be fraught with alarming dangers.

Bharadvāja speaks highly of Viśvakarmā, Chāyāpuruṣa, Manu, Maya, etc., who have made their magnificent contribution in the field of aeronautics. It appears that they profusely made use of electricity in flying the aeroplanes. *Bṛhad Vimāna Śāstra* also describes in detail the metals employed in the manufacturing of the planes in ancient times.

Aerial Routes (transit routes): Knowledge of the atmosphere was an integral part of ancient Indian aviation science. This need was felt even by ancient Indian scientists of aeronautics. An exclusive section, though brief, has been dedicated to discussion on aerial routes and eddies or Āvartas *(Bṛhadvimāna Śāstra, 1.3-4)*. Bodhnanda has described five aerial or transit routes in his commentary on the *Bṛhadviāman Śāstra,* sūtra no. 1.3:

यथारहस्यविज्ञानं पूर्वसूत्रे निरूपितम् ।
पंचावर्तस्वरूपं च तथैवास्मिन्निरूप्यते ॥1 ॥

yathārahasyavijñānaṁ pūrvasūtre nirūpitam l
paṁchāvartasvarūpaṁ cha tathaivāsminnirūpyate ||1 ||

ऐतेनोभयविज्ञानादेव यन्तृत्वतामियात् ।
इतिसूत्रद्वयविचारात्सिद्धं भवति घ्रुवम् ॥ 2 ॥

aitenobhayavijñānādeva yantṛtvatāmiyāt /
itisūtradvayavichārātsiddhaṁ bhavati ghruvam // 2 //

पंचावर्तविचारस्तु शौनकोक्तप्रकारतः ।
रेखादिपंचमार्गानुसारादत्र प्रकीर्त्यते ॥ 3 ॥

paṁchāvartavichārastu śaunakoktaprakārataḥ /
rekhādipaṁchamārgānusārādatra prakīrtyate // 3 //

रेखापथो मण्डलश्च कक्ष्यशक्तिस्तथैव च ।
केन्द्रेश्चेति विमानानां मार्गाः खे पंचधा स्मृता ॥ 4 ॥

rekhāpatho maṇḍalaścha kakṣayaśaktistathaiva cha /
kendreścheti vimānānāṁ mārgāḥ khe paṁchadhā smṛtā // 4

Five types of aerial routes are-

Rekhā patha (Troposphere, along with the Tropopause): It was known as Śaktyāvarta or whirlpool of Energy.

Maṇḍala patha (Stratosphere): It was vātāvarta or whirlpool of winds.

Kakṣyā patha (Mesosphere): It was Kiraṇāvarta, or whirlpool, made by solar rays.

Śakti patha (Thermosphere): or Śaktyāvarta, i.e. whirlpool of cold Currents.

Kendra patha (Van Allen Belts): It was known as Gharṣaṇāvarta or whirlpool, caused by the collision of solar rays and the earth's magnetosphere.

There were military planes besides luxury and commercial ones. They were equipped with all necessary protective features like wireless, telephone, television, magnetic indicators, Barometer, and X-ray-like machines.

Besides, some instruments could preserve the plane from torrential rains, cyclones, and scorching solar heat.

Military planes were also provided with destructive missiles, which enveloped the enemy plane in pitch darkness when operated. Some instruments similar to microscopes and telescopes were used by which they could visualise the minutest things most guardedly hidden in the enemy's camp and detect the operations of enemies. There were some machines similar to our modern X-Ray but immensely more powerful. These were known as *Guhāgarbha ādarśa yantra*. Some explosive bombs were used, which could spread centrifugal volcanic fire on enemy planes' sides. Some detonative bombs produced hear-shocking and ear-rending sounds that would scare away the enemy forces. Some bombs were capable of blowing away ranges of mountains. It would not be an exaggeration to say that some things like atom bombs and hydrogen bombs were well-known to ancient Indians.

The speed of planes ranged from 1,000 miles per hour to 20,000 miles per hour. That is why the planes were manufactured from lighter materials, the properties of which are detailed exhaustively in *Lohatattva Prakaraṇa* of *Bṛhad Vimāna Śāstra*. They had studied hundreds of metals unknown to us and had studied magnetism, chemistry, sound, light, electricity, air-conditioning, aeronautics, aerostatics, and nuclear physics. It should be noted that sage Bharadvāja is also the author of a highly scientific work named Aṁśubodhini Śāstra, which exhaustively deals with the properties of Solar Rays.

10

High Technology in Ancient India

Several experiments in the field chemistry, metallurgy, and material science described in Sanskrit scientific texts like (a) *Vimāna Śāstra* and (b) *Aṁśu Bodhini*, both ascribed to Maharṣi Bharadvāja, NS (c) *Kṛtakavajra Nirṇaya,* have been carried out with the help of Birla Science Centre Hyderabad, Birla Institute, BHEL-CTI, DMRL and NGRI, and the investigations prove that Sanskrit literature of India had registered a highly developed scientific-technological informations.

Based on the study and investigation performed till now by Dr. CSR Prabhu, Director General (Retd.) National Informatics Centre, Government of India, Ministry of Information Technology, carried out the following twenty formulae for new materials consisting of special alloys, ceramics and glasses have been deciphered and some of them were produced based on the formulae from the ancient Sanskrit texts of *Vimāna Śāstra (quoting Lohatantra), Aṁśu Bodhinī, Kṛtaka Vajra Nirṇaya, etc.* Here, it may be indicated that the below-cited formulae are not available in modern times, thereby hinting at an ancient historical origin.

1. *Tamogarbha Loha:* A lead alloy capable of light absorption has already been produced in the laboratory, light in weight, black in colour, and found to be acid-

resistant. It displayed a high absorption level for laser light (from red Ruby laser - as observed by Prof. Robert Anderson of San Jose State University during his visit to India in December 1991). Some chemicals and other properties were found to be unique - patentable new alloys. A laboratory test done in 1996 in the Physics Department of Osmania University, Hyderabad, indicated laser absorption characteristics up to 79% of incident light from a laser. (This alloy was used in 'Tamo Yantra' in the Vimāna Śāstra for absorption of light escaping from a photochemical reaction, which resulted in light absorption, thereby generating 'darkness').

2. *Pañca Loha:* A copper-alloy which is (not the well-known highly malleable and also Pañcaloha for highly corrosion resistant making idols)" to salt (NaCl) has already been produced and characterised to possess:

a) Golden Yellow Colour (described in the Sanskrit text as *'Hema Varṇam'* or golden colour).

b) Corrosion resistance to moisture and salt water (displayed two weight loss of only about 0.00335 mg/dm/day in 3% NaCl solution).

c) High machinability and microstructure analysis were found to be a single-phase alloy with high malleability (described in Sanskrit as 'm,dulam' or 'soft').

d) Characteristics, composition, and properties were not listed in ASM Reference (1988) and, therefore, patentable new alloy.

3. *Arāra Tāmra:* A copper-alloy zinc, lead, and iron of light absorption have already been produced and characterised to possess:

a) Golden Yellow to reddish tinge (described in

Sanskrit text as (Hema Varnam' or golden colour).

b) Brittle, light, and hard-on microstructure analysis was found to be a two-phase alloy.

c) Very hard (Young's modulus 16.9) (described in Sanskrit text as `Dridham' or `strong').

d) Characteristics, composition, and properties were not listed in ASM Reference (1988) and, therefore, patentable new alloy.

4. *Chapalā grāhaka:* A fine porcelain type of ceramic has already been produced and characterised to be resistant to all acids and alkalis.

5. *Chapalā grāhaka* (glass): A soft glass of low-temperature melt has already been produced and characterised to be resistant to acids and alkalis.

The Refractive index was found to be 1.614. (highest known among soft glasses made at low temperatures).

6. *Ravi Śakti:* A special glass concentrating (visible) light darpana (glass)" energy in sunlight has already been produced, and the study of optical properties is not yet done.

7. *Uṣṇa Śakti:* A special glass for Apakarṣaṇa concentrating the heat darpana (glass) energy in sunlight has been fully deciphered to be produced in the laboratory.

8. *Badhira Loha:* A soundproof alloy has been fully deciphered to be produced in the laboratory.

9. *Vidyut darpaṇa:* A special glass that can neutralise electrical discharges as lightning has been fully deciphered to be produced.

10. ***Rāja Loha:*** A high-heat-absorbing alloy used for dies of various flying crafts has been fully deciphered to be produced in the laboratory.

11. ***Rudanti Mani:*** A particular material has been fully deciphered to be produced in the laboratory.

12. ***Rutika Mani:*** A particular material has been fully deciphered and is to be produced in the laboratory.

13. ***Abhra Mṛd Darpaṇa:*** A special mica glass has been fully deciphered and is to be produced in the laboratory.

14. ***Sunda Mṛt Kaca:*** A special glass has been fully deciphered and is to be produced in the laboratory.

15. ***Pingala Ādarśa:*** A special glass has been fully deciphered and is to be produced in the laboratory.

16. ***Somānka Loha:*** A special alloy has been fully deciphered and is to be produced in the laboratory.

17. ***Ravi Śakti Apakarṣaṇa:*** A special glass with solar Darpaṇa (heat collecting properties) has been fully deciphered and is to be produced in the laboratory.

18. ***Hatakasya Loha:*** A Copper alloy with a golden appearance has been fully deciphered and is to be produced in the laboratory.

19. ***Vāta Stambhana Loha:*** A copper iron, the lead alloy has been Fully deciphered and is to be produced in the laboratory.

20. ***Ghaṇemdashārava Loha:*** An alloy with a high sensitivity to different sounds.

While informing the press on July 18, 1991, of the success in making these metals, Dr B.B. Siddharth, Director, Birla Science Centre, Hyderabad, said that in making these metals, various medicinal leaves, gum, barks of trees, etc., are also used. That is why some unique qualities are developed in the metals while the production cost is less. He further said that if the country's policymakers contemplate the manufacture of the various metals described in the book and how to accumulate the necessary things, then it will be suitable for the country's future development.

The news of the above press conference was released by the news agency Vārtā. It was published on July 19 in Nai Duniya, MP Chronicle, and many other newspapers nationwide. Similarly, Dr. Maheshwar Sheron of the Chemistry Department of IIT, Mumbai, tried to make some things described in the book. These were Cumbakmaṇi, used in the Guhāgarbha yantra and can capture a reflection. Pāragandhika drava - This is a type of acid which is used in the guhāgarbha and is used with a cumbakmaṇi.

Similarly, there is a description of the various kinds of metals and mirrors in Sage Bharadwaj's Anśu-bodhini. Dr. N.G. Dongre, Reader in the Harishchandra PG College, Varanasi, has undertaken a project with the cooperation of the Indian National Science Academy. The project was named 'The Study of Various Materials Described in Anśu-bodhini of Maharshi Bharadvāja'.

Under the project, he tried to make a mirror as described by Sage Bharadwaj at the National Metallurgical Lab, Jamshedpur, with the Director, P. Ramachandra Rao, the Vice-Chancellor of the Benaras Hindu University. He successfully manufactured a special kind of glass called prakāśa stambhana bhid loha. The speciality of this glass is that it absorbs visible light and allows only infrared rays to pass through it.

It has been made of kachar loha-silica bhuca-akra surmitr-adikshar- lime ayaskakānt (lodestone) ruruk (deer bone) ash, as per the process laid down in An "u-bodhini. The speciality of prakaśa stambhan bhid loha is that it is entirely non-hygroscopic. Infra-red hygroscopic mirrors lose their polish and lustre in water vapour or humidity and become useless. These days, CaF2 is highly hygroscopic. Therefore, one has to be extra cautious while using these machines, although a study of the Prakash stambhan bhid loha has proved that it works best in the infra-red range of two to five microns (μ), where $1\mu = 10\text{-}4$ cm, and that it can be used without worrying about the moisture in the atmosphere.

Hence, we can say that the truth behind some of the experiments carried out in one of the chapters of Sage Bharadvāja's book makes us believe that the others must also be true and that aeronautical and other sciences were not just part of fancied imagination in ancient times, but were a fact. More and more research on these ancient scientific texts by courageous researchers will prove the authenticity of the advancement of science and technology in ancient India.

Low-cost Protein-rich food from Grass

In addition to the above materials, a formula for producing a protein-rich food extract (powder) from specific Indian grasses is also deciphered—the Central Food Tech. Research Institute, Hyderabad, has certified that the powder extracted from the specified grasses shows about 13% protein content. Other tests from nutrition and medical angles are to be taken up. This activity aims to produce low-cost protein-rich food products (such as powder, biscuits, malt, etc.) based on this formula.

Agni Stambhana or Fire Resistance

In addition to materials produced or deciphered as above, a technique for preventing and resisting fire and burning is also developed. Two techniques/solutions have been developed for Agni Stambhana or fire resistance:

a) for preventing the burning of inflammable objects like paper, cloth, and wood

b) for preventing the burning of the human body.

It has been noted that in both the above cases, the fire will not be allowed to be caught (or burning to start) even after continuous exposure to flame for up to 30 seconds. (Normally, fire catches any inflammable material within 0.5 seconds, and any moist material dampened with this liquid does not catch fire and can also be used to put off the fire or escape unburnt in fire even after prolonged exposure to flame).

Anāhāra or Avoiding food

A recipe or a particular type of biscuit has been developed. When consumed up to 50 g, this biscuit can help overcome hunger and skip a meal for about 3-4 hours. Upon medical trial, it was found very useful in obesity and diabetic cases.

The Vedic Ion Engine

The National Aeronautics and Space Administration planned a rendezvous with Haley's Comet in the 1980s.

They planned to use a space probe powered by an Ion Engine. This engine used a stream of high-velocity electrified particles instead of a blast of hot gases. The theory of the Ion Engine has been credited to Robert Goddard, long recognised as the father of Liquid-fuel Rocketry. It is claimed that in 1906, long before Goddard launched his first modern rocket, his imagination had conceived the idea of an Ion rocket; however, in light of new evidence, the story could be entirely different. In 1895, on a beach in Mumbai (Bombay, Maharashtra, India), Shivkar Bapuji Talpade, a Sanskrit scholar, proved that heavier-than-air flight was possible. This demonstration was attended by eminent citizens, including, among others, His Highness Maharaja Sayajirao Gaekwad of Baroda and Mr. Justice Ranade, and was reported in "The Kesari", a leading Marathi daily newspaper. Readers might note this occurred eight years before the Wright Brothers' "First" flight at Kitty Hawk, North Carolina, USA.

An even more astonishing feature of Talpade's craft was the power source he used- an Ion Engine. Certain verses in the 10th chapter of the Ṛgveda refer to the Art of Flight. The great Ṛṣi Bharadvāja has written a commentary on this in his book "Yantra Sarvasva" (Science of Machines). Bharadvāja elucidates the mechanism that provides the impulse needed for propulsion. It combines eight sub-assemblies and uses the interaction of principally Solar energy and Mercury. Talpade put his knowledge of Sanskrit at the disposal of his creative intellect and constructed an aircraft according to the description given in the *Ṛgveda*. It is reported that this flying machine gained an altitude of 1500 ft. most aptly, he called his aircraft the

"Marutsakhā," friend of the wind.

The engine, now being developed for future use by NASA, by some strange coincidence, also uses Mercury bombardment units powered by Solar cells. Interestingly, the impulse is generated in seven stages. The Mercury propellant is first vaporised and fed into the thruster discharge chamber, ionised, converted into plasma by a combination with electrons, broken down electrically, and then accelerated through small openings in a screen to pass out of the engine at velocities between 20,000 and 50,000 meters per second.

Although minute details of the Vedic engine would be available only after significant research, the resemblance of the "modern" engine to it is indisputable. The Ion Engine developed by NASA can produce, at best, about one pound of thrust - which is virtually useless for lifting an object of any practical mass off the earth. On the other hand, Talpade's engine was entirely capable of lifting his aircraft 1500 feet into the air over 100 years ago.

Several vital considerations emerge from the foregoing discussion. First, Wilbur and Orville Wright were not the pioneers of modern flight. Secondly, not only had the idea of an Ion Engine been conceived long before Dr. Goddard but it had also been materialised in the form of Talpade's *Marutsakhā Aircraft*. I do not wish to denigrate those inventors whose contributions are invaluable, but I think it is now time to review the history of science and recognise the achievements of the previous civilisations. The question of the exaltation or diminution of any country or civilisation's contribution does not arise. My only contention is that if scientific thought began in Vedic civilisation earlier than in the West, we should not ignore that fact in our narration of the history of science.

Courtesy Ancient Skies, Bi-monthly published by the Ancient Astronaut Society, 192 St. Johns Ave. Highland Park, Illinois 60035 USA.

11
Electrical Engineering

Electricity is said to be the most important achievement of modern Science. Here, it may be pointed out that the Vedic seers also invented electricity. In the Vedas, it has been mentioned by the name Indra. The word *Vidyuta* was also used for the same. We also find the mention in *Agastya Saṅhitā* (Indian Prince's Library) regarding the electricity generation method and constructing electrical batteries. The relevant reference can be quoted under:

संस्थाप्य मृण्मयेपात्रे ताम्रपात्रं सुसंस्कृतम् ।
छादयेच्छिखिग्रीवेण चाद्राभिः काष्ठपांशुभिः ॥
दस्तालेष्टो विधात् पारदाच्छादितस्ततः ।
संयोगाज्जायते तेजो मित्रावरुणसंज्ञितम् ॥
अनेनजलभंगोऽस्ति प्राणोदानेषु वायुषु ।
एवं शतानां कुम्भानां संयोगकार्यकृत्स्मृतः ॥

saṁsthāpya mṛṇmayepātre tāmrapātraṁ susaṁskṛtam /
chhādayechchhikhigrīveṇa chādrābhiḥ kāṣṭhapāṁśubhiḥ //
dastāleṣṭo vidhāt pāradāchchhāditastataḥ /
saṁyogājjāyate tejo mitrāvaruṇasaṁjñitam //
anenajalabhaṁgo'sti prāṇodāneṣu vāyuṣu /
ēvaṁ śatānāṁ kumbhānāṁ saṁyogakāryakṛtsmṛtaḥ //

"Place a well-cleaned copper plate in an earthenware vessel. Cover it first with copper sulfate (*shikhigriva*) and then moist sawdust (*kāṣṭhapānsu*). After that, put a mercury-amalgamated -zinc (*pārad ācchādita dast*) sheet on top of an energy known by the twin name of Mitra-Varuna (cathode and anode). This current will split water into Prana-vāyu (oxygen) and Udāna-vayu (hydrogen). A chain of one hundred jars gives a very active and effective force to make a battery bank (kārya kṛt)."

Further, it is stated that

वायुबन्धकवस्त्रेण निबद्धो यानमस्तके ।
उदानः स्वलघुत्वेन बिर्भिर्त-आकाश-यानकम् ।

vāyubandhakavastreṇa nibaddho yānamastake |
udānaḥ svalaghutvena birbhirta-ākāśa-yānakam |

"That is to say this energy can be used in extracting hydrogen (udāna-vāyu) from water and then the same, being lighter than air can be used in the operation of an aeroplane."

Apart from this, the *Atharvaveda* (2.5.12) refers to two kinds of electricity, i.e. positive and negative, and its friendly and destructive use. According to ÿgveda (1.16.5), the electricity is hidden in the water, and when it comes out, it spreads light and provides energy. The *Ṛgveda* (1.85.5; 188.1) describes the use of electricity in weapons and telegraphy. The word *Vidyut* is used to denote electricity in Vedas. Electricity can be generated from energy (*RV.* 1.45.5). Electricity protects people and should be used as destructive energy against wicked persons and enemies with the help of weapons, which work on electricity (*RV.* 1.86.9). A scientist who knows the nature of Time and all the properties and characteristics of electricity can accomplish his/her work very fast (*RV.* 1.95.8).

The *Atharvaveda*, apart from many other significant

aspects of science like medicinal plants, healing practices, marriage rites, and the healing properties of water, Supreme Consciousness also considers 'electricity'.

Here, it would not be insignificant to highlight the research work done by Dr. Ashok Tiwari, an electrical engineer from M.P. State Electricity Board, Jabalpur, on some aspects of 'Electrical Energy' registered in *Atharvaveda.*

According to him, electricity is one of the most exciting topics in the *Atharvaveda.* It discusses a detailed description of valuable applications that harness and utilize this immense energy source. Some of the applications described include essential specifications for a control system that harnesses the intense power of electricity for use as a deadly weapon, utilization of hydroelectric power for manufacturing, and the fission properties of electricity.

It will be fully apparent from the following mantras that Electrical energy and its properties were fully understood by Vedic sages. It was used in everyday technological applications as much (if not more), as it is today. It is also clear that Vedic scientists knew far more about the properties of electricity than us. It was not until very recently that modern-day scientists discovered and began researching the potential usage of electrical energy in military science and related areas.

What these verses demonstrate to us is the fact that Vedic Society was in a highly advanced state of technological advancement and fully utilized electricity in innumerable productive applications.

According   to   the   descriptions   found   in the *Atharvaveda*, electrical energy can be utilized in many practical applications such as the creation of powerful engines, illumination, agricultural machinery, hydroelectric power plants, manufacturing plants, biomedical engineering, extraction of medicines, etc., and thus serve to enhance the daily life of people greatly. For instance,

The *A V.* (20.7.2) describes the inherent properties of electrical energy. It refers to 'positive' and 'negative currents'.

नव यो नवतिं पुरो बिभेद बाह्वोऽजसा । अहिं च वृत्रहावधीत् ॥

nava yo navatiṁ puro bibheda bāhvo'jasā l ahiṁ cha vṛtrahāvadhīt ll

"Electricity, which breaks, by the energy of its arms (i.e. positive and negative currents) the 99 cities called bhogas in the Vedic terminology (so-called elements known to modern scientists), destroys the cloud, which makes clouds precipitate, the source of energy and power."

The *A V.* (20.7.3) refers to various valuable applications of electricity. The mention of horse-powered driving machines is a direct reference to electronically powered vehicles like automobiles, aircraft, etc. In our times, even prototypes of electric cars are a very recent development. In the present era, electric car prototypes were created because gasoline fuel is a perishable resource and pollutes the environment. Thousands of years ago, Vedic seers were aware of these dangers and prescribed using electric engines in their vehicles as one of their primary modes of transport and for motors, etc., for other types of machinery. This verse also clearly mentions that electricity was used to light up buildings, as we do today. In addition, there is a reference to electrically powered agricultural machines, which helped in the manufacture and processing of crops from the fields.

स नः इन्द्रः शिवः सखाश्वावद् गोमत् यवमत् । उरुधारेव दोहते ॥

sa naḥ indraḥ śivaḥ sakhāśvāvad gomat yavamat /
urudhāreva dohate //

"That very electric power may be our peaceful friend, providing us with the horsepower to drive our machines, light to lighten our houses, and power to produce grains in the fields. Let it bring prosperity and well-being to us by flowing into numerous currents."

The *AV.* (20.7.4) refers to the application of electricity to devise technology for manufacturing different things. The picture that emerges from these *mantras* is that of a highly advanced culture that utilised superlative technology yet maintained enormous respect and reverence for the ecosystem and the natural environment. Not only this, but they had analysed the properties and laws of all these natural energies and thus comprehended perfectly the best way to utilise them.

इन्द्र क्रतुविद् सुतं सोमं हर्यं पुरुष्टुत । पिबा वृषस्व तातृपिम् ॥

indra kratuvid sutaṁ somaṁ haryaṁ puruṣṭuta / pibā
vṛṣasva tātṛpim //

"Let electricity be applied to create different technologies by those well versed in manufacturing. Let it keep safe and satiate all our desires."

The *AV.* (20.31.1) mentions that electric current is propelled by two speedily moving forces of attraction and repulsion, which are powerful like the thunderbolt, pleasant and praiseworthy.

ता वज्रिणं मन्दिनं स्तोम्यं मद इन्द्र रथे बहतो हर्यता हरी ।
पुरूण्यस्मै सवनानि हर्यत इन्द्राय सोमा हरयो दधन्विरे ॥

tā vajriṇaṁ mandinaṁ stomyaṁ mada indra rathe bahato haryatā harī |
purūṇyasmai savanāni haryata indrāya somā harayo dadhanvire ||

"Those two speedily moving forces of attraction and repulsion propel the electric current, powerful like the thunderbolt, pleasant and praiseworthy, in this pleasant plane or car. Manifolds are the generating powers for the refulgent electricity borne by speedy moving *somas* - electric charges."

The *AV.* (20.31.2) discusses the principles of electromagnetic force used to generate motion. In addition, the reference to a liquid fuel propellant informs us that high-speed and intense electrical power was used to produce favourable results like high-speed vehicular modes of transportation and efficient manufacturing processes. For example, high-speed centrifugal force is used to separate genetic materials in Biomedical labs today. The combination of speed and power is certainly desirable to increase the efficiency of manufacturing plants..

अरं कामाय हरयो दधन्विरे स्थिराय हिन्वन्हरयो हरी तुरा ।
अर्वद्भिर्यो हरिभिर्जोषमीयते सो अस्य कामं हरिवन्तमानशे ॥

araṁ kāmāya harayo dadhanvire sthirāya hinvanharayo harī turā |

arvadbhiryo haribhirjoṣamīyate so asya kāmaṁ
harivantamānaśe ||

The above-mentioned speedy forces of two kinds set in motion strong currents capable of maintaining steady progress in attaining one's objective in plenty. Whatever complex is attained by these fast-moving horsepowers is enough to achieve the beautiful objective of his, the manufacturer.

The *AV* (20.15.2) acknowledges the immense power of electricity and provides a deep understanding of its intensity. Electricity does have the power of striking through any element. The portion about productive works depending on waters flowing with speed is an apparent reference to harnessing hydroelectric power. It is evident from this verse that Vedic society was well aware of methods that harnessed electricity power through systems like hydroelectric power plants. Once harnessed, they incorporated the force to implement manufacturing concerns and power machinery.

अध ते विश्वमनु हासदिष्टय आपो निम्नेव सवना हविष्मतः ।
यम्पर्वते न समशीत हर्यत इन्द्रस्य वज्रः श्रथिता हिरण्ययः ॥

adha te viśvamanu hāsadiṣṭaya āpo nimneva savanā
haviṣmataḥ |
yamparvate na samaśīta haryata indrasya vajraḥ śnathitā
hiraṇyayaḥ ||

"Just as all productive works of the manufacturer depend upon waters flowing down with speed, so do all his desired objects depend upon you (Electricity), as any cloud or mountain cannot obstruct its powerful striking force in the way. It smashes all impediments with its radiant energy."

AV. (20.15.3) explains that electricity was researched and explored extensively during the Vedic age. Skilled electrical engineers used to devise specific utilities and controls for electric power. Illumination was undoubtedly one of its widely used applications. From the last line, we can infer that electronic devices were used as hearing aids and magnifiers. Sophisticated systems for controlling electricity and measuring electricity were in place, and electronic panels that aided in controlling other forms of energy were also used.

अस्मै भीमाय नमसा समध्वर उषो न शुभ्र आ भरा पनीयसे ।
यस्य धाम श्रवसे नामेन्द्रियं ज्योतिरकारि हरितो नायसे ॥

asmai bhīmāya namasā samadhvara uṣo na śubhra ā bharā paṇīyase /
yasya dhāma śravase nāmendriyaṁ jyotirakāri harito nāyase //

"O well-versed engineer make use of this terrible electric power fit to be utilized for practical purposes by controlling it, for non-violent, brilliant light like the dawn. It has the potential to help to hear, control energy, and spread light in all quarters."

Now, let us turn to the fantastic descriptions of electricity being utilised as a weapon in military combat. Vedic seers were aware of this potentially devastating aspect of electricity being utilised as a weapon. Only in the past decade or so scientific research has begun concentrating on the potential use of electricity as a deadly weapon. Electrical bombs can be utilised to destroy all sorts of vital equipment and inflict massive damage. Topics like the EMP effect and the consequent possibilities of using electrical weapons were rudimentary information to Vedic people, whereas we have yet to begin our explorations on this topic.

The Electro-magnetic Pulse (EMP) effect was first observed during the early testing of high-altitude airburst nuclear weapons. The effect is characterised by producing a very short (hundreds of nanoseconds) but intense electro-magnetic pulse, which propagates away from its source with ever-diminishing intensity, governed by the theory of electromagnetism. The Electro-Magnetic Pulse is, in effect, an electro-magnetic shock wave.

This energy pulse produces a powerful electro-magnetic field, particularly within the vicinity of the weapon burst. The field can be sufficiently strong to produce short-lived transient voltages of thousands of Volts (i.e. kilovolts) on exposed electrical conductors, such as wires, or conductive tracks on printed circuit boards, where exposed.

This aspect of the EMP effect is of tremendous military significance, as it can result in irreversible damage to a wide range of electrical and electronic equipment, particularly computers and radio or radar receivers. Subject to the electro-magnetic hardness of the electronics, a measure of the equipment's resilience to this effect, and the intensity of the field produced by the weapon, the equipment can be irreversibly damaged or, in effect destroyed. The damage inflicted would be the same as that of striking the equipment with huge lightning bolts.

Computers used in data processing systems, communications systems, displays, and industrial control applications, including road and rail signalling, and those embedded in military equipment, such as signal processors, electronic flight controls, and digital engine control systems, are especially vulnerable to the EMP effect.

What amazes one is that Vedic seers were aware of the EMP effect and composed vivid descriptions of the usage of electric weapons as demonstrated below through an excerpt from *A V.* (20.15.6):

त्वं तमिन्द्र पर्वतं महामुरुं वज्रेण वज्रिन्पर्वशश्चकर्तिथ।
अवासृजो निवृताः सर्तवा अपः सत्रा विश्वं दधिषे केवलं सहः ॥

tvaṁ tamindra parvataṁ mahāmuruṁ vajreṇa vajrinparvaśaśchakartitha ǀ
avāsṛjo nivṛtāḥ sartavā apaḥ satrā viśvaṁ dadhiṣe kevalaṁ sahaḥ ǁ

"Just as the thundering electricity reduces the vast cloud to nothing by its thunderbolt, so do you, O King, equipped with piercing weapons like the thunderbolt, smash into pieces the vast armies of the enemy, consisting of various units, by your striking power like the thunderbolt. Just as the waters of the cloud released by the electricity fall and flow over the earth, similarly, the well-equipped armies of the enemy, being subdued by the king's might, are duly regulated by him. Honestly do you alone, O King, hold all the power to subdue the foes."

The inference is quite apparent to weapons utilizing electricity. "Piercing weapons like the thunderbolt" is a clear pointer to surges of exceedingly high voltage. The lethal electric weapons are used to counter various units of the army. This is another clue, for as discussed above, the EMP effect can be used to advantage for several targets ranging from computers to communication systems. Electricity was employed as one of the primary weapons in military combat during the Vedic era.

In *AV.* 20.38.5, we find a mention that electricity has the striking power of a deadly weapon. There can be no more significant proof of the deployment of electrical weaponry during the Vedic era than this.

इन्द्र इद्धर्योः सचा सम्मिश्ल आ वचोयुजा। इन्द्रो वज्री हिरण्ययः ॥

indra iddharyoḥ sachā sammiśla ā vachoyujā / indro vajrī hiraṇyayaḥ //

'Electricity is well mixed up with *Prāṇa* (positively charged ions) and *Apāna* (negatively charged ions), the two horsepowers, yoked to the power of speech. Electric power has the striking power of a deadly weapon and is full of brilliance.'

AV. (20. 30. 1) mentions the dual nature of electrical energy. It can be destructive, as in the EMP effect, and it can be protective if utilized in constructive technology.

प्र ते महे विदथे शंसिषं हरी प्र ते वनुषो हर्यतं मदम्।
घृतं न यो हरिभिश्चारु सेचत आ त्वा विशन्तु हरिवर्पसं गिरः ॥

pra te mahe vidathe śaṁsiṣaṁ harī pra te vanuṣo haryataṁ madam /
ghṛtaṁ na yo haribhiśchāru sechata ā tvā viśantu harivarpasaṁ giraḥ //

"O electricity, I fully praise thy two forces of protection and destruction in this great universe, which is a great sacrificial place or battlefield of life. I highly cherish your beautiful exhilaration, destroying the evil forces of the enemy. You shower various forms of fortunes through your blessing powers of speedy action, like waters from the clouds. Let all praises find their abode in you of charming splendor."

AV. (20.30.3) describes different sorts of electrical weapons. The "beautiful horse of iron of high speed" apparently refers to some metallic car/aircraft that can reach incredibly high speeds. The horsepower for the engine for a craft or automobile of this sort was provided by electricity. The "shining arrow" can mean a potent ballistic missile loaded with an electric warhead. The missile was superfast and covered vast distances; it was one of the primary weapons that could be used to target the central command and control centre of the enemy. Here, we find a reference telling us that many advanced weapons utilizing electricity had been deployed in this era.

सो अस्य वज्रो हरितो य आयसो हरिर्निकामो हरिरा गभस्त्योः ।
द्युम्नी सुशिप्रो हरिमन्युसायक इन्दे नि रूपा हरिता मिमिक्षिरे ॥

so asya vajro harito ya āyaso harirnikāmo harirā
gabhastyoḥ I
dyumnī suśipro harimanyusāyaka inde ni rūpā haritā
mimikṣire II

"Here is the blue-green-coloured thunderbolt of iron of the king. There is also the beautiful horse of iron at high speed. Here is also the horsepower of the rays of electricity. There is also the shining arrow, capable of destroying the enemy's pride and having very high speed. In short, many kinds of weapons have been made through electric power for the king."

AV. (20. 30. 4) describes a missile that seems to

generate immense power and would be exceptionally destructive. The electric weapons used by Vedic society may have been equivalent in destructive power to nuclear weapons or even more lethal. They may also have been used for preliminary strikes before the actual use of nuclear weapons. A conventional electronic combat campaign, or intensive electronic combat operations, would initially concentrate on saturating the opponent's electronic defences, denying information, and inflicting maximum attrition upon electronic assets. The mass application of electromagnetic bombs in the opening phase of an electronic battle would allow much faster attainment of command as it would destroy electronic assets at a much faster rate than possible with conventional means. After this phase, it would be child's play to destroy the enemy completely.

दिवि न केतुरधि धायि हर्यतो विव्यचद्वज्रो हरितो न रंह्य ।
तुददहिं हरिशिप्रो य आयसः सहस्रशोका अभवद्धरिंभरः ॥

divi na keturadhi dhāyi haryato vivyachadvajro harito na raṁhya /
tudadahiṁ hariśipro ya āyasaḥ sahasraśokā abhavaddhariṁbharaḥ //

"Like a radiant spot, it is well placed in the heavens, then with a high speed, the destructive missile, made of iron, possessing the speed of electric power, crushing the serpent natured enemy, becomes lit up with thousands of lights and loaded with the destructive ray of various kinds."

AV. (20.21.7) refers to a remotely controlled electronic weapon system. It directly points to unleashing a tremendously high voltage surge that can blast into the defensive electrical equipment belonging to the enemy. The last sentence describes the remote initiation of the bombing sequence while the controller is stationed at a safe distance.

युधा युधमुप धेदेषि धृष्णुया पुरा पुरं समिदं हंस्योजसा।
नम्या यदिन्द्र सख्या परावति निबर्हयो नमुचिं नाम मायिनम्॥

yudhā yudhamupa dhedeṣi dhṛṣṇuyā purā puraṁ samidaṁ haṁsyojasā I
namyā yadindra sakhyā parāvati nibarhayo namuchiṁ nāma māyinam II

"O mighty King, thou can quickly get at the striking power of the enemy by thy overwhelming striking force. Being well-entrenched in thy sheltered place of defence, thou can break the enemy's defences to smithers. Thoroughly crush the deceitful enemy, unfit to be left alive, through thy faithful ally, although stationed at a distance."

AV. (20.21.8) refers to electricity as overcoming all resistances and obstructions in the way. Any trained electrical engineer must know the complex laws and limitations governing electrical energy. There is evident respect for electricity, as it is an incredible power. Transport infrastructure can also be destroyed with the use of electromagnetic bombs. Railway and road signalling systems, where automated, are most vulnerable to electromagnetic attacks on their control centres. Significantly, most modern automobiles and trucks use electronic ignition systems, which are known to be vulnerable to electromagnetic weapons effects.

Modern land warfare doctrine emphasises mobility, and manoeuvre warfare methods are typical for contemporary land warfare. Coordination and control are essential to the successful conduct of manoeuvre operations, and this provides another opportunity to apply electromagnetic weapons. Communications and command sites are critical elements in the structure of such a land army, and these have concentrated communications and computer equipment. Therefore if the enemy is attacked with electromagnetic weapons, the entire command and control of land operations would be completely disrupted. Employing tactics like launching an initial attack with electromagnetic weapons would create a maximum of confusion, and if followed by an all-out attack with conventional weapons, would completely obliterate the enemy.

त्वं करंजमुत पर्णय बधीस्तेजिष्ठयातिथिग्वस्य वर्तनी ।
त्वं शता वंगृदस्याभिनत्पुरोऽनानुदः परिषूता ऋजिष्वना ॥

tvaṁ karaṁjamuta parṇaya badhīstejiṣṭhayātithigvasya vartanī |
tvaṁ śatā vaṁgṛdasyābhinatpuro'nānudaḥ pariṣūtā ṛjiṣvanā ||

"O mighty electricity, you kill the violent enemy, equipped with speedy means of communication like cars or airships, a hindrance in the way of people worthy of respect, cows, or land by your consuming and splendorous power. You shatter the innumerable forts of the adversary who obstructs your communications or breaks your regulations and does not pay tribute to you established by straightforward negotiations."

AV. (20.21.9) talks about the single transmitter of high quality to control the entire emission and circularly polarized emission.

त्वमेतां जनराज्ञो द्विर्दशाबन्धुना सुश्रवसोपजग्मुषः ।
षष्टिं सहस्रा नवतिं नव श्रुतो नि चक्रेण रथ्या दुष्पदावृणक् ॥

tvametāṁ janarājño dvirdaśābandhunā
suśravasopajagmuṣaḥ /
ṣaṣṭiṁ sahasrā navatiṁ nava śruto ni chakreṇa rathyā
duṣpadāvṛṇak //

"O electricity, you can by your circular motion like the wheel of a chariot, which is too powerful to be checked, well keep under control all these 20 basic elements, 6099 organic, and inorganic bodies, by a single transmitter of high quality, with no other force to help it."

In technological terms, the coupling is the means to create an electric connection between two electric circuits by having a part common to both. One of the mechanisms that can be exploited to improve coupling is the polarization of an electric weapon's emission. Polarization means orienting the emission in a particular direction. If we assume that the orientations of possible coupling apertures and resonances in the target set are random in relation to the weapon's antenna orientation, a linearly polarized emission would only exploit half of the opportunities available. A circularly polarized emission would exploit all coupling opportunities. A single transmitter of superior quality can control the entire emission.

AV. (20.21.10) apparently refers to a defence mechanism created with electricity, which protects both the commander and his entire system of wireless communications. It may refer to some shield that makes things unsusceptible to the emission. The advanced and deadly missiles of unlimited speed are controlled by electronic control panels.

 त्वमाविथ सुश्रुवसं तवोतिभिस्तव त्रामभिरिन्द्र तूर्वयाणम् ।
त्वस्मै कुत्समतिथिग्वमायुं महे राज्ञे यूने अरन्धनाय ॥

tvamāvitha suśruvasaṁ tavotibhistava trāmabhirindra
tūrvayāṇam /
tvasmai kutsamatithigvamāyuṁ mahe rājñe yūne
arandhanāya //

"O electricity, you keep in safety this good listening set, by your means of safety and protect the commander, with speedy mobile forces by your strong means of defense. You control the sharp weapons, equipped with the striking power of limitless time and speed for this great, youthful king."

Other electronic devices and electrical equipment would also be destroyed by the EMP effect. Telecommunications equipment can be highly vulnerable, due to the presence of lengthy copper cables between devices. Receivers of all varieties are particularly sensitive to EMP, as highly sensitive miniature high-frequency transistors and diodes in such equipment are easily destroyed by exposure to high voltage electrical transients. Therefore radar and electronic warfare equipment, satellite, microwave, UHF, VHF, HF, and low band communications equipment and television equipment would all be destroyed by the EMP effect.

AV. (20.37.4) is a simple testament to the intense power and utterly destructive force of electrical weapons.

त्वं नृभिर्नृमणो देववीतौ भूरीणि वृत्रा हर्यश्व हंसि ।
त्वं नि दस्युं चुमुरिं धुनिं चास्वापयो दभीतये सुहन्तु ॥

tvaṁ nṛbhirnṛmaṇo devavītau bhūrīṇi vṛtrā haryaśva

haṁsi /

*tvaṁ ni dasyuṁ chumuriṁ dhuniṁ chāsvāpayo dabhītaye
suhantu* //

"O electrical currents of high voltage, safely carried by electric wires, you kill many enemies in the war waged by learned persons or through the help of natural forces. To keep all the evil forces under control, you, being well-equipped with suitable means of destruction, ultimately lay down to lasting sleep (death) the evil forces that rob and harass the general public."

There are many more such references to electrical energy in the *Atharvaveda.* The discovery of such amazing truths in the Vedas clearly undermines the very foundation of our smug assertions that Civilisation and scientific advancement proceed linearly. It is high time for us to thoroughly analyse and learn from the invaluable resources that our forefathers have left us in the form of the divine Vedas and other profound Vedic literature.

12

Metallurgy

Most ancient metals such as iron, gold, silver, lead, tin, brass, and zinc were invented in India during the Vedic period itself. All these metals were naturally available in India.

Loha or ayas was a term commonly applied to metal. So, iron was called kṛṣṇāyas or kṛṣṇa loha, and copper was called lohitāyas/raktāyas or lohita loha/rakta loha. The Atharvaveda (11.3.7) makes the mention of iron as śyāma-aya and copper as lohita.

श्याममयोऽस्य मांयानि लोहितमस्य लोहितम्।

śyāmamayo'sya māṁyāni lohitamasya lohitam ǀ

"Black iron His flesh, copper His blood"

We, however, also find the use of ayas in the meaning of iron alone. For example, the Atharvaveda (5.28.1) makes a mention of harita (gold), rajata (silver), and ayas (loha).

हरिते त्रीणि रजते त्रीण्यसि त्रीणि तपसा विष्ठितानि।

harite trīṇi rajate trīṇyasi trīṇi tapasā viṣṭhitāni ǀ

"In a later period, the term loha or ayas came to stand for iron alone."

Besides gold, silver, iron, copper, trapu (tin), sisa (lead), and bronze have also been mentioned in the Vedas. For instance,

The Atharvaveda (11.3.8) mentions trapu (tin).

त्रपुभस्म हरितं वर्णः पुष्करमस्य गन्धः ।

trapubhasma haritaṁ varṇaḥ puṣkaramasya gandhaḥ ।

"Tin is His ashes, gold his colour, the blue lotus flower His scent."

Vājasaneyī Saṁhitā of Yajurveda (18.13) sīsa (lead) and trapu (tin) in addition to gold, iron, and copper.

हिरण्यं च मे अयश्च मे श्यामं च मे सीसं च मे लोहश्च मे त्रपुश्च मे यज्ञेन कल्पन्ताम् ।

hiraṇyaṁ cha me ayaścha me śyāmaṁ cha me sīsaṁ cha me lohaścha me trapuścha me yajñena kalpantām ।

"Extract gold, bronze, iron, lead, copper, and tin for my prosperity."

In the Vedic period, metals like gold and silver were known and used for ornaments of various descriptions. The Ṛgveda (1.64.4) makes a mention of various types of ornaments to cover various body parts.

चित्रैरंजिभिर्वपुषे व्यंजते वक्षःसु रुक्माँ अधि येतिरे शुभे ।
अंसेष्वेषां नि मिमृक्षुऋष्टयः साकं जज्ञिरे स्वधया दिवो नरः ॥4 ॥

chitrairaṁjibhirvapuṣe vyaṁjate vakṣaḥsu rukmāṁ adhi yetire śubhe ।
aṁseṣveṣāṁ ni mimṛkṣuṛrṣṭayaḥ sākaṁ jajñire svadhayā divo naraḥ ॥4 ॥

"They decorate their persons with various ornaments; they have put garlands for elegance and brilliance between their arms, i.e. on the neck; also, the weapons are borne upon their shoulders, and with them and their own strength, they emerged as leaders."

In the *Ṛgveda* (1.122.14), the person who used to

wear ear-rings was called Hiraṇyakarṇa (having golden ears), and who used to wear the garland of the rosary of bead was called Maṇigriva.

हिरण्यकर्णं मणिग्रीवमर्णस्तन्नो विश्वे वरिवस्यन्तु देवाः ।

अर्यो गिरः सद्य आ जग्मुषीरोस्राश्चाकन्तूभयेष्वस्मे ॥14 ॥

hiraṇyakarṇaṁ maṇigrīvamarṇastanno viśve varivasyantu devāḥ ।

aryo giraḥ sadya ā jagmuṣīrosrāśchākantūbhayeṣvasme ॥14 ॥

"May all the divine powers favour a person decorated with golden earrings and a jewel necklace. May the venerable (the company of the deities) be propitiated by the praises issuing (from the mouth of the worshipper); may our offerings be acceptable to them, and (may they be pleased) with both (our praises and offerings)."

During the Vedic period, the jeweller was called Maṇikāra (VS. 30.7), and the goldsmith was known as Hiraṇyakāra (VS. 30.17). Mentioning both words reveals that during the Vedic period, ornaments made of gold and silver were also worked with various types of gems.

Gold was often called by the name 'yellow metal' and silver by the name 'white metal'. Similarly, iron was called by the name 'black metal', and copper was called by the name 'red metal.'

In the Atharvaveda (1.15.2;4), we find the mention of bullets made of lead. For instance,

सीसायाध्याह वरुणः सीसायाग्निरुपावति ।

सीसं म इन्द्रः प्रायच्छत् तदंग यातुचातनम् ॥ 1.15.2

sīsāyādhyāha varuṇaḥ sīsāyāgnirupāvati ।

sīsaṁ ma indraḥ prāyachchhat tadaṁga yātuchātanam ॥

"The Varuṇa (king) advises the use of lead, the Agni (commander) protects the subjects through the use of lead bullets, the Indra (commander-in-chief) has given me lead bullets for safety. O dear, lead verily repels the fiends."

यदि नो गां हंसि यद्यश्वं यदि पूरुषम् ।
तं त्वा सीसेन विध्यामो यथा नोऽसो अवीरहा ॥ 1.15.4

yadi no gāṁ haṁsi yadyaśvaṁ yadi pūruṣam /
taṁ tvā sīsena vidhyāmo yathā no'so avīrahā // 1.15.4

"If you kill our cow, human being, or a horse, we pierce you with this lead bullet so that you may not slay our men and animals."

The lead was known to Vedic people by several names like sisa, sisak, naga, dhatuvisha, uraga, etc. I am trying to remember the mention of ingot lead (sisa) used as money, but we came across a reference in the Yajurveda (19.80) to lead ingots used as money. The mantra reads under:

सीसेन तन्त्रं मनसा मनीषिणः ऊर्णासूत्रेण कवयो वयन्ति ।

sīsena tantraṁ manasā manīṣiṇaḥ ūrṇāsūtreṇa kavayo
vayanti /

"Wise people apply their intelligence to manufacture coins with lead and clothes with woollen thread."

Similarly, we find mention of lead ingots in the *Atharvaveda* (12.2.53), where sisa (lead) is mentioned as money. The mantra goes like this:

सीसं क्रव्यादपि चन्द्रं त आहु ।

sīsaṁ kravyādapi chandraṁ ta āhu /

"Your money is made of sisa (lead)."

In Chhāndogya Upaniṣad (4.17.7) also states that gold can be bound using borax and silver using gold and tin using silver and lead using tin and iron using lead and wood using iron or leather.

लवणेन सुवर्ण संदध्यात् सुवर्णेन रजतं रजतेन वपु वपुणा सीसं सीसेन लोहं लाहेन दारु दारु चर्मणा।

lavaṇena suvarṇaṁ saṁdadhyāt suvarṇena rajataṁ rajatena vapu vapuṇā sīsaṁ sīsena lohaṁ lāhena dāru dāru charmaṇā /

Yājñavalkya Smṛti (Vya. 178) informs about the loss of a quantity of metals in percentage when melted in the fire. Accordingly, gold does not lose any quantity when melted in a fire.

Silver lost 2% per cent of its quantity, copper 5%, iron 10%, zinc, and led incurs a loss of 8 % of their quantity.

अग्नौ सुवर्णमक्षीणं द्विपले रजते शते। अष्टौ त्रपुषि सीसे च ताम्रे पंच दशायसि॥

agnau suvarṇamakṣīṇaṁ dvipale rajate śate / aṣṭau trapuṣi sīse cha tāmre paṁcha daśāyasi //

We also come across specimens of pottery and metallic sheets from the sites of Harappa and Mohanjodaro. Bronze, copper, iron, lead, gold, and silver made axes, daggers, knives, spears, arrowheads, swords, drills, metal mirrors, ingots, eating, cooking and storage utensils. Of 324 objects excavated from these archaeological sites, 184 are made of pure copper (Lahiri, 1995) and four archaeological copper processing kilns are reported at Harappa, Lothal, and Mohanjo-daro's sites. (Agarwal, 2000: 40)

Greek Historians, such as Herodotus, Ktesias, Arrian, and Megasthenes, have documented the use of metals in India (Bigwood, 1995).

The Painted Grey Ware of Gangetic Valley was manufactured around 600 BC (Dikshit, 1969, p. 3). Two glass bangles found at Hastinapur, near New Delhi, are from 1100-300 BC (Dikshit, 1969, p. 3). The bangles were found to have been made from soda-lime-silicate Glass with ornamental colouring from iron. Similar bangles of green colour were also found elsewhere in northern India (Dikshit, 1969, p. 3). Glass bangles are ornamented jewellery worn by Indian females daily. Beads of green Glass with cuprous oxide colouring were found in Taxila in 700-600 BC (Dikshit, 1969, p. 4).

In his book 'On Marvelous Things Heard', Aristotle talks about the excellent quality of Indian copper, which could be distinguished from gold. This mention was made perhaps to acknowledge the high quality of Indian bronze made from copper, indistinguishable from gold due to the mixing of different metals or minerals (Kumar, 2014 p. 221).

In addition to the above, we find Ayurvedic texts like Charaka, Suśruta (38.62; 46.326-328), and Aṣṭāṅga Hṛdaya dealing with various types of metals and herbo-metallic medicines. In the medieval ages, we have Rasaratna Samuccaya, Rasāraṇavam, Yajñavalkya Samṛti, and some other texts dealing elaborately with the field of Metallurgy.

Unfortunately, very little material is now available to enable us to present a connected narrative of the metallurgical skill of the ancient Indians. Many vital links still need to be included; we shall here try to put together only a few, which we have been able to recover.

Magasthenes says that the Indians were 'well skilled in the arts'. According to the Greek writer, the soil, too, has "underground numerous veins of all sorts of metals, for it contains much gold and silver, and copper and iron in no small quantity and even tin and other metals, which are employed in making articles of use and ornament, as well as the implements and accoutrements of war." (Ray, 1903: p. 153)

Based on the above-mentioned ancient books dealing with metallurgy, we discuss in detail some of the essential aspects of ancient Indian metallurgy.

Mining

Mining was undertaken at a large scale in ancient India. Ancient Indians were well aware of the significance and contribution of mining to the state economy. According to Kauṭilya (16th Century B.C) (2.12), mines are a source of treasury, and the treasury gives power or authority to govern. A state can be governed by treasury and power.

आकरमभवः कोशः कोशाद् दण्डः प्रजायते ।
पृथिवी कोशदण्डाभ्यां प्राप्यते कोशभूषणा ।

ākaramabhavaḥ kośaḥ kośād daṇḍaḥ prajāyate |
pṛthivī kośadaṇḍābhyāṁ prāpyate kośabhūṣaṇā |

Kauṭilya has devoted two chapters (2.12-13) to mining. He also appointed a superintendent of mining called Ākarādhyakṣa to oversee the mining activities in the country. According to him (2.12), the superintendent of mines should possess the knowledge of the science dealing with copper and other minerals

(Sulbádhátusástra), experienced in the art of distillation and condensation of mercury (rasapáka) and of testing gems (Maṇirāga), aided by experts in mineralogy and equipped with mining labourers and necessary instruments.

आकराध्यक्षः शुल्बधातुशास्त्ररसपाकमणिरागज्ञस्तज्ज्ञसखो वां तज्जातकर्मकरोपकरणसम्पन्नः किट्टमूषाङ्गारभस्मलिङ्गं वाकरं भूतपूर्वमभूतपूर्व बा भूमिप्रस्तररसधातुमत्यर्थवर्णगौरवमुग्रगन्धरसं परीक्षेत ।

ākarādhyakṣaḥ   śulbadhātuśāstrarasapākamaṇirāga-
jñastajjñasakho   bāṁ   tajjātakarmakaropakaraṇasampannaḥ
kiṭṭamūṣāṅgārabhasmaliṅgaṁ   vākaraṁ   bhūtapūrva-
mabhūtapūrva   bā   bhūmiprastararasadhātumaty-
arthavarṇagauravamugragandharasaṁ parīkṣeta /

"Possessed knowledge of the science dealing with copper and other minerals (Sulbádhátusástra), experienced in the art of distillation and condensation of mercury (rasapáka) and of testing gems (Maṇirāga), aided by experts in mineralogy and equipped with mining labourers and necessary instruments, the superintendent of mines shall examine mines which, on account of their containing mineral excrement (kiṭṭa), crucibles (mūṣa), charcoal (aṅgāra), and ashes (bhasma), may appear to have been once exploited or which may be newly discovered on plains or mountain-slopes possessing mineral ores, the richness of which can be ascertained by weight, depth of colour, piercing smell, and taste."

Extraction of pure metals

c). Extraction of pure metals without sparks of fire, bubbles, crystallization, and cracking sounds to get pure metal alloys like a jewel.

d). Gold mining with the help of ants.

e). Various types of bronzes, brass, bell metal, lead, tin, various classes of iron, etc., along with the process of their extraction and purification.

The technology of extraction and culture of gems

The scientists and experts of mining science, as registered in Vedic and post-Vedic Sanskrit literature, developed technology of extraction of metals and gems and an art of culture of gems. The technocrats and scientists were conversant with the science of (metal) veins in the earth and metallurgy, the art of smelting and art of colouring gems, or having the assistance of experts here and fully equipped with workers skilled in the work and with implements, should inspect an old mine by the marks of dross, crucibles coal and ashes, or a new mine where there are ores on the earth, in rocks or in liquid form, with excessive colour and heaviness and with a strong smell and taste.

The technocrats and mining scientists were so skilled that they developed simple tests to examine the presence of metal ores and gems. From the liquids that flow inside a hole, a cave, a tableland, and at the foot, a rock-cut cave, or a secret dugout in the mountains, the mining scientists and experts confirm which type of metal is present in these samples. If the colour of these samples is of the colour of the jambu, or the mango or the palm fruit of, a cross-section of ripe turmeric or jaggery or of orpiment or red arsenic or honey vermilion or white lotus or of the feather of a parrot or of a peacock at the same time if they have water and plants of the same colour around them and that are viscous, clear and heavy, the gold will be available there.

When thrown in the water, they spread on the surface like oil and absorb mud and dirt; they can transmit copper and silver up to one hundred times their own weight.

At the same time, similar in appearance but full of a strong smell and taste is the sign of bitumen.

Ores in earth or rocks, which are yellow or copper-coloured or reddish yellow, which, when broken, show blue lines or are of the colour of the mudga or masa-bean or krasara, which are variegated with spots or lumps as of curds, which are of colour of turmeric or myrobalan or lotus leaf or moss or liver or spleen or saffron, which, when broken, show lines, spot or svastikas of fine sand which are possessed of pebbles and are lustrous, which, when heated, do not break and yield plenty of foam and smoke, are gold-ores, to be used for insertion, as transmuters of copper and silver. [43]

In the case of all ores, when there is an increase in heaviness, there is an increase in metal content.

When the metal content is not clear, the Maurayan scientists confirm that those samples of ores are impure or dim in the interior flow in a pure form when infused in strong urine and caustic, when formed into lumps with a paste of the rājavṛkas, the banyan tree, the pilu, cow's bile, pigment and the urine and dung of the buffalo, the donkey and the young camel, either mixed with this during the boiling or smeared with this paste. The impure ore became pure, and impurities were quickly removed.

The insertion in the ore of the bulbous roots of the kadali and the vajra along with the caustic which is made up of the ashes of barley husks(barley), masa-bean (bean-talk), sesamum (sesamum-talks), Palāśa (wood of palāśa) and pilu (wood of pilu tree) was used for the productivity of the softness in the metal. The insertion in the boiling ore, the milk of the cow and the goat is also a reagent to produce softness in the metal.

The scientists and experts during the period of imperial Mauryas invented that if honey and liquorice, goat's milk with sesamum oil, mixed with ghee, jaggery, and fermenting stuff, together with the kandali with only three infusions in the melted ores, the product will become so soft metal that it may break into a hundred-thousand fold.

To produce hard metal, they inserted the powder of cow teeth and horns of a cow.

After collecting ores, they physically examined the ore, and based on physical properties, they distinguished whether the ore was of gold, silver, copper, iron, etc. We have already discussed the ores of gold and silver. Now, we discuss other metals. The ore from rocks or a region of the earth, which is heavy, unctuous, soft and tawny, green, reddish, or red, is copper ore. [49]

The ore which is crow-black or of the colour of the dove or yellow pigment or studded with white lines and smelling like raw flesh is lead ore.[50]

The grey ore, like saline earth or of the colour of a baked lump of earth, is tin ore. [51]

The one made mostly of smooth stones is whitish red, and the sinduvāra flower's colour is iron-ore.[52]

The ore which is of the colour of *kākāṇḍa* (crow's egg) or birch leaf is *vaikṛntaka-ore*.[53]

The ore clear, smooth, lustrous, possessed of sound, cold, stiff, and light colour is gem-ore.[54]

So, we see a sequence between the physical examination methods of the metals in Vedic and post-Vedic Sanskrit literature. Side by side, the scientific and technological development registered in Vedic and post-Vedic Sanskrit literature was advanced, and the scientists and technologists of that age developed the technique of the culture, extraction, and collection of metals and gems.

Identification of the ores of various metals

According to Kauṭilya (2.12), the following is the identification of the ores of various metals.

Ores of Gold: 1. Liquids which ooze out from pits, eaves, slopes, or deep excavations of well-known mountains; which have the colour of the fruit of rose-apple (jambu), of mango, and of fan palm; which are as yellow as ripe turmeric, sulphurate of arsenic (*haritāla*), honeycomb, and vermilion; which are as resplendent as the petals of a lotus, or the feathers of a parrot or a peacock; which are adjacent to (any mass of) water or shrubs of similar colour; and which are greasy (*chikkaṇa*), transparent (*viśada*), and very heavy are ores of gold (*kāñchanika*).

पर्वतानामभिज्ञातोद्देशानां   बिलगुहोपत्यकालयनगूढखातेष्वन्तः   प्रस्यन्दिनो जम्बूचूततालफलपक   हरिद्राभेदेहरितालक्षौद्रहिङ्गुलक   पुण्डरीकशुकमयूरपत्रवर्णाः सवर्णोदकौषधिपर्यंताश्चिक्कणा विशदा भारिकाश्च रसाः काञ्चनिकाः ।

parvatānāmabhijñātoddeśānām

*bilaguhopatyakālayanagūḍhakhāteṣvantaḥ-   prasyandino
jambūchūtatālaphalapaka
haridrābhedeharitālakṣaudrahiṅgulaka   puṇḍa-
rīkaśukamayūrapatravarṇāḥ
savarṇodakauṣadhiparyanāśchikkṇā viśadā bhārikāśścha rasāḥ
kāñchanikāḥ ⁄*

Those ores which are obtained from plains or slopes of mountains; which are either yellow or as red as copper or reddish yellow; which are disjoined and marked with blue lines; which have the colour of black beans (māṣa, Phaseolus Radiatus), green beans (mudga, Phaseolus Mungo), and sesamum; which are marked with spots like a drop of curd and resplendent as turmeric, yellow myrobalan, petals of a lotus, aquatic plant, the liver or the spleen; which possess a sandy layer within them and are marked with figures of a circle or a svastika; which contain globular masses (sagulikā); and which, when roasted do not split, but emit much foam and smoke are the ores of gold (suvarṇa dhātavaḥ) and are used to form amalgams with copper or silver (pratīvāpārthās-tāmra--rūpya-vedhanāḥ).

पीतकास्ताम्रकास्ताम्रपीतका वा भूमिप्रस्तरधातवो भिन्ना नीलराजीमन्तो मुद्गमाषकृसरवर्णा वा दधिविन्दुपिण्डचित्रा हरिद्राहरीतकीपद्मपत्रशैवल यकृत्प्लीहानवद्यवर्णा भिन्नाश्चुञ्चुवालुकालेखाबिन्दुस्वस्तिकवन्तः सगुलिका अर्चिष्मन्तस्ताप्यमाना न भिद्यन्ते बहुफेनधूमाश्च सुवर्णधातवः प्रतीवापार्थास्ताम्ररूप्यवेधनाः ⁄

*pītakāstāmrakāstāmrapītakā   vā   bhūmiprastaradhātavo
bhinnā   nīla-   rājīmanto   mudgmāṣkṛsaravarṇā   vā
dadhivindupiṇḍachitrā   haridrāharītakī-   padmapatraśaivala
yakṛtplīhānavadyavarṇā   bhināśchuñchuvālukāle
khābindusva- stikavantaḥ sagulikā archiṣmantastāpyamānā
na bhidyante bahuphenadhū- māścha suvarṇadhātavaḥ
pratīvāpārthāstāmrarūpyavedhanāḥ ⁄*

Ores of Silver: 1. Likewise, liquids which, when dropped on water, spread like oil to which dirt and filth adhere and which amalgamate themselves more than cent per cent (satadupari veddharah) are ores of copper or silver.

अप्सु निष्ठ्यूतास्तैलवद्विसर्पिणः पङ्कमलग्राहिणश्च ताम्ररूप्ययोः शतादुपरि बेद्धारः ।

apsu nisthyūtāstailavadvisarpiṇaḥ paṅkamalagrāhiṇaścha tāmrarūpyayoḥ śatādupari bedvāraḥ ।

Those ores which have the colour of a conch shell, camphor, alum, butter, a pigeon, turtle dove, Vimalaka (a kind of precious stone), or the neck of a peacock; which are as resplendent as opal (sasyaka), agate (gomédaka), cane-sugar (guda), and granulated sugar (matsyandika) which has the colour of the flower of kovidára (Bauhinia Variegata), of lotus, of patali (Bignonia Suaveolens), of kalaya (a kind of phraseolus), of kshauma (flax), and of atasi (Dinuin Usitatissimum); which may be in combination with lead or iron (anjana); which smell like raw meat, are disjoined grey or blackish white, and are marked with lines or spots; and which, when roasted, do not split, but emit much foam and smoke are silver ores.

शङ्खकर्पूरस्फटिकनवनीत   कपोतपारावतविमलकमयूरग्रीवावर्णाः सस्यकगोमेदकगुडमत्स्यण्डिकावर्णाः   कोविदारपद्मपाटलीकलायक्षौ   मातसी पुष्पवर्णाः ससीसाः साञ्जना विस्रा भिन्नाः श्वेताभाः कृष्णाः कृष्णाभाः श्वेताः सर्वे वा लेखाविन्दुचित्रा मृदवो ध्मायमाना न स्फुटन्ति बहुफेनधूमाश्च रूप्यधातवः ।

śaṅkhakarpūrasphaṭikanavanīta kapotapārāvatavimalaka-mayūragrīvāvarṇāḥ sasyakagomedakaguḍamatsyandikāvarṇāḥ kovidārapadmapāṭ-alīkalāyakṣau mātasī puṣpavarṇāḥ sasīsāḥ sāñjanā visrā bhinnāḥ śvetābhāḥ kṛṣṇāḥ kṛṣṇāmāḥ śvetāḥ sarve vā lekhāvinduchitrā mṛdavo dhmāyamānā na sphu-

ṭanti bahuphenadhūmāścha rūpyadhātavaḥ /

Kauṭilya (16th Century B.C) also informs that the heavier the ores, the greater the quantity of metal in them (*satvavriddhiḥ*).

सर्वधातूनां गौरववृद्धौ सत्त्ववृद्धिः ।

sarvadhātūnāṁ gauravavṛddhau sattvadṛddhiḥ /

Test of Metals

By colouration in fire: According to Rasārṇava (Vol.1, p.68), copper yields blue flame; that of tin is pigeon-coloured; that of lead is pale tinted; that of iron is tawny; that of peacock ore (sāyaska) is red. (Ray, 1909: Vol. 1, p. 68)

By melting: According to Rasārṇava, pure metal is that which, when melted in the crucible, does not give off spark nor bubbles, nor surts, nor emit any sound, nor shows any lines on the surface, but is tranquil like a gem or say shows the sign of tranquil fusion. (Ray, 1909: Vol. 1, p. 68)

Locations for mining

Locations of copper mines: Those ores that are obtained from plains or slopes of mountains and which are heavy, greasy, soft, tawny, green, dark, bluish-yellow (harita), pale-red, or red are ores of copper.

भारिकः स्निग्धो मृदुश्च प्रस्तरधातुर्भूमिभागो वा पिङ्गलो हरितः पाटलो लोहितो वा ताम्रधातुः ।

bhārikaḥ snigdho mṛduścha prastaradhāturbhūmibhāgo vā piṅgalo haritaḥ pāṭalo lohito vā tāmradhātuḥ /

Locations of lead mines: Those ores which have the colour of kākamechaka (Solanum Indica), pigeon, or cow"s bile, and which are marked with white lines and smell like raw meat are the ores of lead.

काकमेचकः कपोतरोचनावर्णः श्वेतराजिनद्धो वा विस्रः सीसधातुः ।

kākamechakaḥ kapotarochanāvarṇaḥ śvetarājinddho vā visraḥ sīsa- dhātuḥ ।

Locations of tin mines: Those ores which are as variegated in colour as saline soil or which have the colour of a burnt lump of earth are the ores of tin.

ऊषरकर्बुरः पङ्कलोष्टवर्णो वा त्रपुधातुः ।

ūṣarakarburaḥ paṅkaloṣṭavarṇo vā trapudhātuḥ ।

Locations of tikṣaṇa (Iron) mines: Those ores which are of orange colour (kurumba), or pale-red (pándurohita), or of the colour of the flower of sinduvára (Vitex Trifolia) are the ores of tīkṣṇa.

कुरुम्बः पाण्डुरोहितः सिन्दुवार पुष्पवर्णो वा तीक्ष्णधातुः ।

kurumbaḥ pāṇḍurohitaḥ sinduvāra puṣpavarṇo vā tīkṣaṇadhātuḥ ।

Locations of brass mines: Those ores which are of the colour of the leaf of kākāṇḍa (Artemisia Indica) or of the leaf or birch are the ores of vaikrintaka.

काकाण्डभुजपत्रवर्णो वा वैकृन्तकधातुः ।

kākāṇḍabhujapatravarṇo vā vaikṛntakadhātuḥ ।

India was the first country to introduce all sorts of metals, specifically brass and zinc. According to modern scholars' estimates, Rajasthan has a zinc ore mine from the second millennium BC. Taxila housed a fourth-century BC brass vase that used zinc metal. Nāgarjuna and others have described the extraction of zinc. During archaeological excavations, several cities scattered all over

India predating the Christian era have been located as having mining activities.

Refining Process of ores and metals

According to Kauṭilya (2.12), the impurities of ores, whether superficial or inseparably combined with them, can be got rid of, and the metal melted when the ores are (chemically) treated with Tikshna urine (mūtra) and alkalies (kshára), and are mixed or smeared over with the mixture of (the powder of) Rajavriksha (Clitoria Ternatea), Vata (Ficus Indica), and Pelu (Carnea Arborea), together with cow's bile and the urine and dung of a buffalo, an ass and an elephant.

तेषामशुद्धा मूढगर्भा वा तीक्ष्णमूत्रक्षारभाविता रराजवृक्षवटपीलु-गोपित्तरोचनमहिषखरकरभमूत्रलण्डपिण्डबबद्धास्तत्प्रतीवापास्त-दवलेपावा विशुद्धाः श्रवन्ति।

teṣāmaśuddhā mūḍhagarbhā vā tīkṣaṇamūtrakṣārabhāvitā rarājavṛkṣavaṭapīlu-gopittarochanamahiṣakharakarabhamūtralaṇḍapiṇḍababaddh āstatpratīvāpāsta-davalepāvā viśuddhāḥ śravanti I

Purification of Gold: We know that gold excavated from different methods is invariably by no means a pure metal. It contains up to 20 per cent silver, copper, iron, lead, bismuth, platinum group metals, and other metals as impurities. Thus, native gold would have different colours depending upon the nature and amount of impurities present. It is logical to assume that the different colours of native gold were a major driving force for the development of the gold refining process. Although evidence of gold refining is available in Vedic texts in an allegory form, we find it in detail in the Arthaśastra of Kauṭilya.

Gold refining was a two-stage process. The first stage was melting impure gold and lead, which removed base metal impurities but not noble metals like silver. The second stage was to heat impure gold sheets with the soil of Sindhu State, which contained salt. The sodium chloride in the soil reacted with silver, and the resulting silver chloride was absorbed into the surrounding soil. This solid-state process involved the diffusion of silver in impure gold and the subsequent formation of silver chloride at the gold-soil interface. Kauṭilya (2.14) describes the process of removing the impurity of gold of whitish colour as:

पाण्डु श्वेतं चाप्राप्तकम् । तद् येनाप्राप्तकं तच्चतुर्गुणेन सीसेन शोधयेत्, सीसान्वयेन भिद्यमानं शुष्कपटलैर्दध्मापयेत्, रूक्षत्वाद् भिद्यमानं तैलगोमये निषेचयेत् ।

pāṇḍu śvetaṁ chāpātakam / tad yenāprāptakaṁ tachchaturguṇena sīsena śodhayet, sīsānvayena bhidyamānaṁ śuṣkapaṭalairmāpayet, rūkṣatvād bhiya-mānaṁ tailagomaye niṣechayet /

" Impure gold is whitish. It shall be fused with lead for four times the quantity of the impurity. When gold is rendered brittle owing to its contamination with lead, it shall be heated with dry cow dung (*suṣkapaṭala*). When it splits into pieces owing to hardness, it shall be drenched (after heating) in oil mixed with cow dung (t*aila gomaye*)."

Gold attained through mines is sometimes found contaminated with lead and assumes a brittle state owing to its contamination with lead. According to Kauṭilya, to remove this type of impurity of gold, it should be heated wound round with cloth (pākapatrāni kritvā) and hammered on a wooden anvil. Or it may be drenched in a mixture made of mushroom and vajrakhaṇḍa (Antiquorum).

आकरोद्गतं सीसान्वयेन भिद्यमानं पाकपत्राणि कृत्वा गण्डि कासु कुट्टयेत् कन्दलीवज्रकन्दकल्के वा निषेचयेत् ।

ākarodgataṁ sīsānvayena bhidyamānaṁ pākapatrāṇi kṛtvā gaṇḍi - -kāsu kuṭṭayet kandalīvajrakandakalke vā niṣechayet /

It is important to note that Kautilya stated that the starting sheet of impure gold must be thin, as this would improve the kinetics of the solid-state refining. Usage of gold in granular form, as was the case at least in part in the Sardis refinery of the Lydian kingdom of Anatolia, would result in lower yield.

Purification of Silver: Impure silver was heated with a lead of one-fourth the quantity of the impurity.

विपर्यये स्फोटनं च दुष्टम् । तत् सीस चतुर्भागोन शोधयेत्।

viparyaye sphoṭanaṁ cha duṣṭam / tat sīsa chaturbhāgona śodhayet /

Purification of Mica: According to Rasaratna Samucchaya (2.17), mica, heated seven times and plunged into Kāñjī (sour gruel) or cow's urine or decoction of Triphalā (chebulic myrobalans) or cow's milk, is freed from all impurities.

प्रतप्तं सप्तवाराणि निक्षिप्तं का/ञ्जिकेऽभ्रकम् ।
निर्दोषं जायते नूनं प्रक्षिप्तं वाऽपि गोजले ॥ 2.16
त्रिफलाक्वथिते चापि गवां दुग्धे विशेषतः ॥ 2.17

*prataptaṁ saptavārāṇi nikṣiptaṁ kāñjake'bhrakam /
nirdoṣaṁ jāyate nūnaṁ prakṣiptaṁ vā'pi gojale // 2.16
triphalākvathite chāpi gavāṁ dugdhe viśeṣataḥ // 2.17*

Softening of metals

In ancient India, softening and hardening technology was also fully developed. As requirements, metals could be softened and hardened.

According to Kauṭilya, metals are rendered soft when they are treated with (the powder of) kandali (mushroom) and vajrakanda, (Antiquorum) together with the ashes of barley, black beans, palāśa (Butea Frondosa), and pelu (Carnea Arborea), or with the milk of both the cow and the sheep. Whatever metal is split into a hundred thousand parts is rendered soft when it is thrice soaked in the mixture made up of honey (madhu), madhuka (Bassia Latifolia), sheep's milk, sesamum oil, clarified butter, jaggery, kiṇva (ferment) and mushroom.

यवमाषतिलपलाशपीलुक्षारैर्गोक्षीराजक्षीरैर्वा

कदलीवज्रकन्दप्रतीवापो मार्दवकरः ।

मधु मधुकमजापयः सतैलं घृतगुडकिण्वयुतं सकन्दलीकम् ।

यदपि शतसहस्रधा विभिन्नं भवति मृदु त्रिभिरेव तन्निषेकैः ॥

yavamāṣatilapalāśapīlukṣārairgokṣīrājakṣīrairvā kadalīvajrakanda- pratīvāpo mārdavakaraḥ /

madhu madhukamajāpayaḥ satailaṁ ghṛtaguḍakiṇvayutaṁ sakandalīkam / yadapi śatasahasradhā vibhinnaṁ bhavati mṛdu tribhireva tanniṣekaiḥ //

Permanent softness (*mṛdustambhana*) is also attained when the metal is treated with the powder of cow's teeth and horn.

गोदन्तश्रृङ्गप्रतीवापो मृदुस्तम्भनः ।

godantaśṛṅgapratīvāpo mṛdustambhanaḥ /

Hardening of metals

Varāhamihira (7th century BC), in his famous treatise the Bṛhatsaṃhitā (50.25-26), while dealing with the chapter on swords (Khaḍagalakṣaṇam), provides a process for the hardening of steel. His process involves chemical techniques as well as heating and quenching. Swords treated with this process did not 'break on stones' or become 'blunt on other instruments'.

Accordingly, a sword dubbed with gingelly oil and then smeared with an unguent prepared with the milky juice of the Arka plant, the cleaned powder of goat's horn and the excreta of doves and mice must be quenched in one of the milk of mare, camel, elephant, deer, or goat mixed with the toddy of palm trees, and afterwards whetted; such a sword will not break on stones.

आर्क पयो हुडुविषाणमपीसमेतं
पारावताखुशकृता च युतः प्रलेपः ।

शस्त्रस्य तैलमथितस्य ततोऽस्य पानं
पश्चाच्छितस्य न शिलासु भवेद्विघातः । २५

ārka payo huḍuviṣāṇamapīsametaṃ
pārāvatākhuśakṛtā cha yutaḥ pralepaḥ |
śastrasya tailamathitasya tato'sya pānaṃ
paśchāchchhitasya na śilāsu bhavedvidhāḥ. 25

Similarly, an iron weapon treated with a day-old drink made of the brunt powder of plantains (or its rib) mixed with butter-milk, and when sharpened properly, will not break on stones nor become blunt on other iron instruments.

थारे कदल्या मथितेन युक्ते दिनोषिते पायितमायसं यत् ।
सम्यक् शितं चाश्मनि नैति भङ्गं न चान्यलोहेष्वपि तस्य कौष्ठ्यम्

thāre kadalyā mathitena yukte dinoṣite pāyitamāyasaṁ yat | samyak śitaṁ chāsmani naiti bhaṅgaṁ na chānyaloheṣvapi tasya kauṣṭyam

Tempering Technology

Tempering technology was also first developed in ancient India. This technology was generally applied to the arrowheads.

According to Vāsiṣṭha Dhanurveda (verses 60-61), an arrowhead is tempered with the root of the reed plant. When the colour of the white reed plant turns yellow after receiving rainwater on Svāti Nakṣatra day, its root becomes poisonous. This root should be procured. The best way to recognise the plant is that it trembles always, even in the absence of wind.

इषुफले शरवंशामूललेपनाद् भवति, तच्चिह्नमेतत्।
यस्मिन् शरवंशसमूहे स्वाति बिन्दुर्निपतति स पीतवर्णो
भवति तस्य मूले विषमुत्पद्यते तन्मूलं ग्राह्यम्।
स च सर्वदा पवनाभावऽपि कम्पते इदमेव तल्लक्ष्मेति ॥ 60-61 ॥

iṣuphale śaravaṁśāmūlalepanād bhavati, tachchihnametat |
yasmin śaravaṁśasamūhe svāti bindurnipatati sa pītavarṇo
bhavati tasya mūle viṣamutpadyate tanmūlaṁ grāhyam |
sa cha sarvadā pavanābhāva'pi kampate idameva
tallakṣameti || 60-61 ||

Long pepper (pippali), saindhava (rock-salt), kuṣṭha (medical plant named Costus Speciosus or arabicus used as a remedy for the disease called takman, in Hindi, this plant is called as Kūṭha) should be ground and pounded in the urine of a cow to prepare a paste. That paste should be smeared over the arrow-head or a weapon, then heated on fire till it becomes blue like a peacock's neck. When it absorbs the entire paste, it must be quenched in water. Such weapons will become the best weapons.

पिप्पली सैन्धवं कुष्ठं गोमुत्रे तु सुपेषयेत्।
अनेन लेपयेच्छसं लिप्तं चाग्नौ प्रतापयेत्॥ 63॥
शिखिग्रीवानुवर्णाभं तप्तपीतं तथौषधम्।
ततस्तु विमलं तोयं पाययेच्छस्त्रमुत्तमम्॥ 64॥

pippalī saindhavaṁ kuṣṭhaṁ gomutre tu supeṣayet /
anena lepayechchhasraṁ liptaṁ chāgnau pratāpayet // 63 //
śikhigrīvānuvarṇābhaṁ taptapītaṁ tathauṣadham /
tatastu vimalaṁ toyaṁ pāyayechchhasramuttamam // 64 //

Sadāśiva Dhanurveda (57) speaks nonetheless in the same terms. Accordingly, long pepper (pippali), saindhava (rock-salt), kuṣṭha (medical plant named Costus Specious or arabicus used as a remedy for the disease called takman, in Hindi, this plant is called as kūṭha) should be ground and pounded by mixing urine of a cow to prepare a paste. That paste should be smeared over the arrow-head or a weapon, then heated on fire till it becomes blue like a peacock's neck. When it absorbs the entire paste, it must be quenched in water. Such weapons will become the best weapons.

पिप्पली सैन्धवं कुष्ठं गोमुत्रेण च पेषयेत्।
अनेन लेपयेच्छसं लिप्तं चाग्नौ प्रतापयेत्॥
अविष्पातं बलाविद्धं पीतमग्नौ तथौषधम्।
ततो निर्व्वापितं लोहं तत्र वेधे विशिष्यते॥ 57॥

pippalī saindhavaṁ kuṣṭhaṁ gomutreṇa cha peṣayet /
anena lepayechchhasraṁ liptaṁ chāgnau pratāpayet //
aviṣpātaṁ balāviddhaṁ pītamagnau tathauṣadham /
tato nirvvāpitaṁ lohaṁ tatra vedhe viśiṣyate // 57 //

Further (Sadāśiva Dhanurveda (58-59) says the paste prepared out of five types of salt soaked in honey and mustard oil is also best for tempering. The arrowhead should be besmeared with this paste and then heated over a fire. When it becomes blue like a peacock's neck, and the whole paste disappears, quench it in the water. It will become the best weapon.

पंचभिर्लवणैः पिष्टैर्मधुसिक्तैः ससर्षपैः ।
एभिः प्रलेपयेच्छस्त्रं लिप्तं चाग्नौ प्रतापयेत् ॥ 58 ॥
शिखिग्रीवानुवर्णाभं तप्तपीतं तथौषधम् ।
ततस्तु विमलं तोयं पाययेच्छस्त्रमुत्तमम् ॥ 59 ॥

pamchabhirlavaṇaiḥ piṣṭairmadhusikttaiḥ sasarṣapaiḥ |
ēbhiḥ pralepayechchhasram liptaṁ chāgnau pratāpayet ||
śikhigrīvānuvarṇābhaṁ taptapītaṁ tathauṣadham |
tatastu vimalaṁ toyaṁ pāyayechchhasramuttamam || 59 ||

According to Sadāśiva Dhanurveda (56) and Vāsiṣṭha Dhanurveda (62), the tempered arrowheads were able to pierce even the unbreakable armours just like a leaf of a tree.

फलस्य पायनं वक्ष्ये दिव्यौषधि विलेपनम् ।
येन दुर्वेध्य वर्म्माणि भेदयेत्तरुपत्रवत् ॥ सदाशिव, 56 ॥
फलस्य पायनं वक्ष्ये दिव्यौषधि विलेपनैः ।
येन दुर्भेद्यवर्म्माणि भेदयेत्तरुपर्णवत् ॥ वासिष्ठ, 62 ॥

phalasya pāyanaṁ vakṣaye divyauṣadhi vilepanam |
yena durvvedhya varmmāṇi bhedayettaru-patravat ||
sadāśiva, 56 ||
phalasya pāyanaṁ vakṣaye divyauṣadhi vilepanaiḥ |
yena durbhedyavarmmāṇi bhedayettaruparṇavat ||
vāsiṣṭha, 62 ||

Scientific methods of physical examination of purity of metals

Experts and skilled persons examined the purity and variety of Gems and metals based on the colour of metals and Gems. Kautilya's Arthaśāstra (2.13) takes note of this method.

The Gold, having a colour of the lotus filament, soft, smooth, not producing a sound and lustrous, was adjudged as the best; the reddish yellow hue was considered to be of middle quality and the red colour was of the lowest quality.

किञ्जल्कवर्णं मृदु स्निग्धमनादि भ्राजिष्णु च श्रेष्ठं, रक्तपीतकं मध्यमं, रक्तमवरं श्रेष्ठानाम् ।

kiñjalkavarṇa mṛdu snigdhamanādi bhrājiṣṇu cha śreṣṭhaṁ, raktapītakaṁ madhyamaṁ, raktamavaraṁ śreṣṭhānām ।

The gold with a light yellow hue was considered the best quality.

The gold with whitish-yellow colour was categorised as impure. The pale yellow and white coloured gold were also said to be impure.

पाण्डुश्वेतं चाप्राप्तकम् ।

pāṇḍutaṁ chāprāptakam ।

Silver, which was white, glossy, and ductile, was described as of the best quality, and that which was of the opposite quality and tended to burst was said to be of bad quality.

तुत्थोद्गतं गौडिकं काम्बुकं चाक्रवालिकं च रूप्यम् श्वेतं।

tutthodgataṁ gauḍikaṁ kāmbukaṁ chākravālikaṁ cha rūpyam । śvetaṁ

स्निग्धं मृदु च श्रेष्ठम् । विपर्यये स्फोटनं च दुष्टम् ।

snigdhaṁ mṛdu cha śreṣṭham । viparyaye sphoṭanaṁ cha duṣṭam ।

The silver with a crest appearing at the top was

classified as impure, clear, lustrous and with a colour of curd was categorised as pure.

उद्गतचूलिकमच्छं भ्राजिष्णु दधिवर्णे च शुद्धम् ।

udgatachūlikamachchham bhrājiṣṇu dadhivarṇe cha śuddham /

Scientific methods of physical examination of purity of gems

A pearl with a big round, without a flat surface, lustrous, white, heave, smooth and perforated at the proper place was adjudged as excellent.

A lentil-shaped, triangular, tortoise-shaped, semi-spherical pearl with a layer, coupled, cut up, rough, spotted, gourd-shaped, dark, blue, and badly perforated was considered defective.

The ruby of the colour of the lotus, the colour of saffron, the colour of Pārijāta -flower, of the morning Sun, that of the beryl, of the blue lotus, water, green bamboo, the parrot's wing, yellow, Cow's urine and of Cow's fat was considered to be of genuine quality.

The Gems, which had a hexagonal square of a round of a flashing colour, having a suitable form, clear, smooth, heavy, lustrous, with lustre inside, were found to be of excellent quality.

The Vimalaka (white-red), green, the Śasyāka (blue), the Añjanamulaka (dark blue), the pittaka (of the colour of Cow's bile, the Sulabhaka (white), the Lohitākṣa (black in the centre and red at the fringe), the Mṛgāsmaka (white and black), the Jyotirakṣa (white blue), the Maleyaka (vermilion-coloured), Ahicchatraka (of a faint red colour), the Kūrpa (white sand grains inside), the Pratikurpa (of the colour of Śikthaka (bees-wax), the Sugandhikurpa (of the colour of the mudga-ben), the Kṣīravaka (milk coloured), the sūkticurṇaka (many coloured), the silapravalaka (coral coloured), the Pulaka (with a black interior), the Śuklapulaka (with a white interior).

These are the subsidiary types of gems.[13]

The rest are not gems but glass crystals.[14]

The gems and rubies with a dull colour and lustre with grains, with a hole in the bloom, broken, badly bored, and covered with scratches, are of the worst quality.[15]

A diamond that is big, heavy, capable of bearing blows, with symmetrical points, capable of scratching a vassal, revolving like a spindle, and brilliantly shining is said to be excellent.[16]

The diamond with points lost, without edges, and defective on one side is imperfect.

In this way, the Mauryan scientists and experts continued and developed the methods and techniques of physical examination of metals and gems of the Vedic period. Colours, shapes, shining and other physical properties with peculiarity were the basis of tests of purity and impurity, due to which it was easy to choose the natural gold, silver, and gem.

Manufacturing of Metals

Kauṭilya (2.12) appointed a superintendent of metals called Lohādhyakṣa to carry on the manufacture of copper, lead, tin, vaikṛntaka (mercury [?]), ārakūta (brass), vṛtta (?); Kaṁsa (bronze or bell-metal), tāla (sulphurate of arsenic), and lodhra (?), and also of commodities (bhāṇda) from them.

लोहाध्यक्षः ताम्रसीसत्रपुवैकृन्तकार कूटवृत्तकं सताललोहकर्मान्तान् कारयेत्, लोहभाण्डव्यवहारं च ।

lohādhyakṣaḥ tāmrasīsatrapuvaikṛntakā ra kūṭavṛttakaṁ satālalohakarmāntān kārayet, lohabhāṇḍavyavahāraṁ cha /

Gold

Nāgārjuna (4th century B.C.), a Buddhist philosopher, alchemist, scientist, and metallurgist par excellence in the age of Śri Yajña Sātakarṇī (404 BC-359 B.C.), was perhaps capable, according to Hiuen Tsang, to convert stone into gold which appears impossible. Also, it is said about the Nāgārjuna that he produced a mountain of gold during famine so that people could buy food grains from foreign lands (Saligrama Krishna, 1985, p. 18). In his famous work on alchemy called ' Rasaratnākara '(1.1), he talks about the secret process of converting silver and copper into gold.

Conversion of sliver into gold:

किमत्र चित्रं यदि राजवर्त्तकं शिरीषपुष्पाग्ररसेन भावितम् ।
सितं सुवर्णंतरुणार्कसन्निभं करोति गुंजाशतमेकगुंजया ॥ 1 ॥

kimatra chitraṁ yadi rājavarttakaṁ śirīṣapuṣpāgrarasena bhāvitam /
sitaṁ suvarṇamtaruṇārkasannbhiṁ karoti guṁjāśatamekaguṁjayā // 1 //

"What wonder is it that the Rājāvartaka (?) digested with the juice of forepart of the Śirīṣa flower (Acacia sirisa) convert silver of the weight of one guñjā into one hundred times its weight of gold of the lustre of the rising sun?"

Further, he discloses (Rasaratnākara, 1.2) that (*pita gandhaka*) yellow sulphur, purified with the extract of palāsa (Butea frondosa), converts silver into gold when roasted thrice over the fire of cow dung cakes.

किमत्र चित्रं यदि पीतगन्धकं पलाशनिर्यासरसेन शोधितः ।
आरण्यकैरुत्पलकैस्तु याचितः कारेति तारं त्रिपिटेन कांचनम् ॥

kimatra chitraṁ yadi pītagandhakaṁ palāśaniryāsarasena
śodhitaḥ /
āraṇyakairutpalakaistu yāchitaḥ kāreti tāraṁ tripiṭena
kāṁchanam //

When darada (cinnabar in English and Singriph in Hindi) is digested several times with the milk of sheep, various acids impart the silver lustre of gold glowing as saffron.

किमत्र चित्रं दरदः सुभावितः पयेन मेष्याः बहुशोऽम्लवर्गैः ।
सितं सुवर्णं बहुधर्मभावितं करोति साक्षाद्वरकुंकुमप्रभवम् ॥

kimatra chitraṁ daradaḥ subhāvitaḥ payena meṣyāḥ
bahuśo'mlavargaiḥ /
sitaṁ suvarṇaṁ bahudharmabhāvitaṁ karoti
sākṣādvarakuṁkumaprabhavam //

Conversion of copper into gold: According to Nāgarjuna, calamine roasted thrice with copper converts the latter into gold.

किमत्र चित्रं रसको रसेन क्रमेणकृत्वाम्बुधरेण रंजितः ।
करोति शुल्वं त्रिपुटेन कांचनम् ॥

kimatra chitraṁ rasako rasena krameṇakṛtvāmbudhareṇa
raṁjitaḥ /
karoti śulvaṁ tripuṭena kāṁchanam //

Varieties and Colour of Gold: Arthśāstra of Kauṭilya (2.13) describes five different varieties of gold with five different hues.

जाम्बूनदं शातकुम्भं हाटकं वैणवं श्रृङ्गिशुक्तिजं, जातरूपं रसविद्धमाकरोद्गतं च सुवर्णम् ।

jāmbūnadaṁ   śātakumbhaṁ   hāṭakaṁ   vainavaṁ śṛṅgiśuktijaṁ,   jātarūpaṁ   rasaviddhamākarodgataṁ   cha suvarṇam

1. Jāmbūnada: It is the product of the river, Jambu emanating from Meru mountain in the Himalayas. Its hue resembles the juice of Jāmuna fruit.

3. Sātakumbha: It is extracted from the mountain of Śatakumba. Its colour is like that of lotus filament (padma kiñjalka)

4. Hāṭaka: It is extracted from the mines. Its colour resembles the yellow colour of kuraṇḍaka flower (yellow kind of barleria. Bot. name Barleria Longifolia) also called yellow amarnath.

5. Vainava: It is the product of the Veṇu mountain and its hue is like that of karṇikāra (kanak champā) flower.

6. Śṛṅgisūktija: It is extracted from gold mines (śṛṅgisūkti) and embued with the colour of manāśśila (red arsenic).

The above varieties of gold may be obtained either pure or amalgamated with mercury or silver or alloyed with other impurities as mine gold (ākarodgata).

Silver

Varieties and colours of Silver: There are four varieties of silver.

1. Tutthodgata: It is extracted from the mountain called Tuttha. It resembles the hu e of Jāti flower (Royal Jasmine. Botanical name 'Jasminum grandiflorum').

2. Gauḍika: It is the product of the country known as Gauḍa or Kāmarūpa (Present-day Assam). Its colour is like that of the flower of 'Tagara' flower (Crape Jasmine. Botanical name 'Tabernamontanna Coronaria).

3. Kāmbuka, It is extracted from the mountain called Kambu. It resembles the hue of Kunda (Jasmine) flower.

4. Chākravālika: It is extracted from the mountain Chakravāla. It also has the colour of Kinda (jasmine) flower.

तुत्थोद्गतं गौडिकं काम्बुकं चाक्रवालिकं च रूप्यम् ।

tutthodgataṁ gauḍikaṁ kāmbukaṁ chākravālikaṁ cha rūpyam /

See Bhaṣya:

रूप्यं चतुर्विधमाह -- तुत्थोद्गतमिति । तुत्थोद्गतं तुत्थपर्वतजं जाति- कुसुमाभं, गौडिकं कामरूपजं तगर कुसुमवर्णे, काम्बुकं कम्बुपर्वतोत्पन्नं, चाकवालिकं चक्रवालाकरोद्धृतं तदुभयं कुन्दपुष्पाभम् इति चतुर्धा रूप्यं श्रेष्ठम् ।

rūpyaṁ chaturvidhamāha-- tutthodgatamiti / tutthodgataṁ tutthaparvatajaṁ jāti- kusumābhaṁ, gauḍikaṁ kāmarūpajaṁ tagara kusumavarṇe, kāmbukaṁ kambuparvatotpannaṁ, chākavālikaṁ chakravālākarodbhūtaṁ tadubhayaṁ kundapuṣpābham iti chaturghā rūpyaṁ śreṣṭham /

Iron

The iron was invented in India during the Vedic period itself. 'Ayas' occurred frequently in the Vedic texts. The term kṛṣṇa-ayas was mainly used to denote iron, while rakta-ayas denoted copper in ancient India. The simultaneous use of kṛṣṇa-ayas and rakta-ayas suggests that iron was a by-product of copper metallurgy. It is also possible that the invention of the iron might have taken place at the hands of copper smelters. Iron was invented indigenously in India, and the independent beginning of iron technology started in India only. The glaring proof is that the diffusion of iron technology took place from India and its migrants to Asia Minor. The Boghaz Keui inscriptions bearing the name of ÿgvedic Gods are ample proof of it. This was why the iron became the monopoly of Hittites and Mitannis outside India.

Iron objects manufactured in India were famous worldwide until the middle of the first millennium BCE. Herodotus (5th century B.C.), the Greek historian, mentions the use of iron-tipped arrowheads by the Indian soldiers in the battle of Thermopylae. Ktesias, in the 4th century B.C., gratefully acknowledged the gift of Indian made sword given to him in the Persian court by the king and his mother. Almost at this very time, Alexander received iron ingots from Northwest India. Arrian mentions the export of Indian iron to the Abyssinian post. The famous Wootz steel earned a special status in the ancient world.

Thus, iron technology in India was not only a consequence of the ingenuity of the skilled metalsmiths here, it gained a special status in the ancient world.

Unfortunately, there is a dearth of sufficient material to present the connected account of ancient India's metallurgical skill and progress. Although many vital links are missing, we tried our best to put together a few facts based on data we could recover.

Megasthenes says that "Indians were well skilled in the arts." According to the Greek writer, the soil, too, has "underground numerous veins of all sorts of metals, for it contains much gold and silver, and copper and iron in no small quantity and even tin and other metals which are employed in marking articles of use and ornaments as well as implements and accoutrements of war." (Ray, 1903: 153)

Zinc

Historically, brass, an alloy of copper and zinc, was known to man much earlier than they could extract zinc from its ore on a large scale. In the early period, zinc was designated as sattva of zinc ore. In the medieval period, it was designated as yashada in Sanskrit. Zinc oxide, known as pushpāñjan, has been referred to in *Charak Saṁhitā*.

Mica

According to the *Rasaratna Samucchaya* (2.4), there are four varieties of mica (Abhraka). 1. Pināka 2. Nāga 3. Maṇḍuka 4. Vajra and each of these are again of four different colours, namely white, red, yellow, and black.

पिनाकं नागमण्डूकं वज्रमित्यभ्रकं मतम् ।
श्वेतादिवर्णभेदेन प्रत्येकं तच्चतुर्विधम् ॥ 4 ॥

pinākaṁ nāgamaṇḍūkaṁ vajramityabhrakaṁ matam /
śvetādivarṇabhedena pratyekaṁ tachchaturvidham // 4 //

Mica with easily easily detached layers is preferred.

सुखनिर्मोच्यपत्रं च तदभ्रं शस्मीरितम् । 2.11

sukhanirmochyapatram cha tadabhram śasmīritam | 2.11

Mica, which is as bright as the moon and has the lustre of the rust of iron, does not take up or combine with mercury. That which has taken up mercury can alone be used with the metals and administered in medicines.

सचन्द्रिकं च किट्टाभ व्योम न ग्रासयेद्रसः ।
गंसितश्च नियोज्योऽसौ लोहे चैव रसायने ॥ 2.12

sachandrikam cha kiṭṭābha vyoma na grāsayedrasaḥ |
gamsitaścha niyojyo'sau lohe chaiva rasāyane || 2.12

Various metals and the process of making different alloys

Historical Evidence

Historical evidence like Jewellery found in Mohenjodaro and the Iron pillar in Delhi, along with the description of Hiuen Tsang, European officers of East India Company, etc., speak loudly about the progress of ancient India in Metallurgy. Hereunder, we cite some marvels of ancient Indian metallurgy.

Iron Pillar of Delhi: The wrought iron pillar of Delhi installed near Qutab-Minar in New Delhi is an example of the metallurgical skill of ancient India. King Chandra constructed and erected the pillar on a hill in the town of Viṣṇupadgiri, as indicated in the three-stanza and long Sanskrit inscription on the pillar. King Chandra of the pillar was Chandragupta II (269 BC-233 BC) of the Gupta dynasty, the son of Samudra Gupta (320 BC-269 BC) by Datta Devi, who ruled Magadha for 36 years. He assumed the title of 'Vikramāditya'. His chief military achievement was his advance to the Arabian Sea through Malva and Gujrat and his subjugation of the peninsula of Saurashtra or Kathiawar, which had been ruled for more than a century by the Śakas known as Satraps. He also vanquished in battle a nation called Bahlika. The elegant, poetic description of the celebrated iron pillar of Mehrauli in Delhi (now known as Qutub Minar) describes his victories.

Iron Pillar of Mehrauli, Delhi

This iron pillar is a marvel of the forger's art of the early Gupta period. The column is a foot and a half thick at the base and twenty-three feet high and is estimated to weigh more than six tons. It is forged in a single piece of pure rustless iron, and more than two thousand years of exposure to every kind of weather has not affected the metal in any way. The current site of the pillar was chosen by Tomar King Anangpal, who installed it on the site of Dhriva stamba (Vaidha yupa yāna) constructed by Varāhāmihira (7th century BC). Between 1192-1199 AD. after the defeat of Prithviraj Chauhan by Qutb-ud-din Aibak, this observatory was defiled, all 27 temples surrounding it were demolished, and it was converted into a Muslim relic by writing ayats of Quran inside it. The ayats show that these were written later, as they could not cover the whole area required to be covered.

In the 18th century, a cannon was fired at the pillar to break it, but it failed (Kumar, 2014, p. 224). This cannonball was most likely used by Nadir Shah, a hungry robber from Persia, in 1739 AD. When he came to Delhi, his army massacred about 30,000 innocent people in just one day, amassed an enormous amount of wealth through loot and arson, and left the area when his hunger for money satiated.

The pillar is about 7.16 m long (23 feet, 6 inches) with a 42.4 cm (16.4 inches) diameter near the bottom and about 30.1 cm (11.8 inches) at the top. The pillar is a solid body with a mechanical yield strength of 23.5 tons per square inch and an ultimate tensile strength of 23.9 tons per square inch (Kumar, 2014, p. 224).

The composition of the wrought iron is as follows: carbon 0.15%, silicon 0.05%, sulfur 0.005%, manganese 0.05%, copper 0.03%, nickel 0.05%, nitrogen 0.02%, phosphorous 0.25%, and 99.4% pure iron (Biswas, 1996, p. 394).

With all the technology available now, it is still impossible to get 99.94% of the purity in iron achieved during that time.

The iron pillar of Delhi is not an isolated example, the Sun Temple of Konark in Odisha was built during the 9th century. There are some 29 iron beams of various dimensions used in its construction. The largest beam is 10.5 metres (35 feet) in length and about 17 cm (7 inches) in width with a square cross-section and is about 2669 kg in weight ray, 1956: 212). Iron nails are used to connect stone pieces. The composition of the iron beams and nails is similar to those of the famed Iron Pillar of Delhi.

Similarly, the Iron Pillar of Dhar is about 1700 years old and has little or no rusting (Kumar, 2014, p. 226). The pillar is 7.3 tons in mass and 42 feet long. Dhar is an old capital city of Malwa, near the well-known city of Indore. The pillar was possibly constructed by King Bhoja (336 AD - 392 AD) and currently lies in three parts in front of La— mosque in Dhar. King Bhoja was well-versed in Metallurgy and wrote a book on metallurgical processes and metal weapons called Yuktikalpataru. His other famous work is Samarāṅgaṇasūtradhāra. The memoirs of Jahangir inform us that Bahadur Shah, a Muslim king, captured the region and decided to move the pillar to Gujrat. While extracting it from the ground, the pillar fell and was broken into two pieces. The pillar has been weathering the monsoons of India for 1700 years and has little rusting. The pillar's surface is coated with a thin, optically dull payer; on top of it is another thick layer of optically bright material (Balasubramaniam, 2002).

In addition to the above, we find an over two-metre-high bronze image of Buddha recovered from Bhagalpur in the Gupta period. Fa-Hsien, a famous Chinese traveller, even saw an over 25 metre-high image of Buddha made of copper, which is untraceable so far (Sigfried J. et al., 1997, p. 388). It is a landmark proving the advancement of metallurgy in ancient India. The science and technology of metallurgy reached its pinnacle in ancient India. It is not an exaggeration to say that India was the father of metallurgy. Nāgārjuna (4th century BC), a Buddhist philosopher, alchemist, scientist, and metallurgist par excellence in the Gupta age, was perhaps capable, according to Hiuen Tsang, of converting stone into gold, which appears impossible.

Not only this, it is said about the Nāgārjuna that he produced a mountain of gold during famine so that people could buy food grains from foreign lands (Saligrama Krishna, 1985, p. 18).

Wootz or Damascus Steel: Indian steel was exported to many countries since very early times and came to be called 'wootz'. It was traded in the form of castings of the size of ice hockey pucks. Persians made swords from wootz, which were erroneously known as Damascus swords (Bigwod, 1995). Like Arabian numerals, the name of Damascus steel came about when Europeans encountered steel in the Middle East around 1192 AD during the War of the Crusades (Kumar, 2014, p. 228). Having glimpsed at the remarkable hardness and strength, they became interested in the secrets of making ultra-high-carbon steel. Europeans at that time were not aware of the fact that this steel was manufactured in India.

Wootz or Damascus steel, as pointed out earlier, is ultra-high carbon steel with high strength. Wootz steel is one and a half times stronger than several worked wrought iron. With proper processing, this steel can be made five times stronger than the most robust wrought iron. Wootz steel was commonly used in making swords and daggers. Thus, we see that no other country in the world could manufacture such steel swords as were made by Indian technicians. They were in great demand in the entire region, from Asia to Europe.

Making wootz steel is a complex process, as even a minute of impurities plays an important role. The temperature and cooling period also decide the quality of steel. Giambattista della Porte (1589 AD) found the importance of temperature in treating wootz. Joseph Maxon (1677 AD) discovered that wootz steel could not be forged above the blood-red heat (Smith, 1982).

Archaeological sites in the Periyar district of Tamil Nadu, dating back to about 250 BC, provide evidence of crucibles used to mix iron and carbon for steel-making (Kumar, 2014, p. 228).

Researchers from Stanford University found that the wootz steel was produced with a slow, constant cooling. When iron and carbon (1.3-1.9%) are heated to 12000C, they reach a molten state, and the slow cooling allows the carbon to diffuse through the iron to form white cementite patterns, called the Damascus steel. The pattern results from the alignment of the ferric ion and carbon (cementite) particles appearing white in the near-black steel matrix. The carbide particles serve the role of strengthening without making the metal brittle (Sherby et al., 1985).

The blade was hardened by heating it to 7270C, allowing a crystal structure change. Iron molecules that were distributed as body-centred ferrite begin to form a face-centred lattice. The blade was then quenched in water (Sherby et al., 1985). It may be reminded here that the heating above the temperature of 8000C would have made the metal brittle.

Thus, one can imagine the precision and high technique used by ancient Indians to manufacture the wootz steel.

The Periplus of the Erythraean Sea (40-70 AD) informs us of steel import to the West. Steel was subjected to customs duty in Rome in the 2nd century

AD. In the Digest of Roman Law, Indian iron (Ferrum Indicum) was on the list of items subject to import duty, as defined by Emperor Justinian (482-565 AD). Similarly, iron was taxed and traded in Alexandria, Egypt, in the early first century AD. Iron was known in those countries as Koos, the corrupt form of the Telgu word 'ukku' or 'urku' used to denote iron. Steel is called wuz in Gujrati language and wooku (or ukku) in the Kannada language of South India. This explains the origin of 'wootz', the Syrian word for steel.

Wootz was also imported into China during the 6th and 7th centuries and was called 'bin iron'. This steel was imported into China as raw material, not as a finished artefact. Several books, such as Wei Shu, written during the 6th century, and Zhou Shu, written during the 7th century, inform about the import of wootz steel and called it fine steel. In 1259, an embassy was sent to the Middle East. Liu Yu, a member of this embassy, mentions 'bin iron' and points out its source as Yindu (Hindu) (Wagner, 2007).

The Arabs called steel Hundwaniy, meaning Indian. The best steel in Persia was called 'fouled-e-Hind', meaning steel of India. Another kind of steel called 'Jawabae Hind', meaning a Hindu answer, was also popular because it could cut a steel sword (Kumar, 2014, p. 231).

13

Textile Engineering

India was rich in raw materials for the principal manufacturing industries. She never developed large-scale industries. All production was made in small-scale industries, according to the needs and requirements of the people and society. Every house was a centre of small-scale industries.

So far as the textile industry was concerned, we find the mention of the following kinds of raw material with the help of which Indians had manufactured different fine and crude varieties of textiles in India. Textile fibres were classified into four categories. 1. Balka, 2. Phāla, 3. Kauṣeya, 4. Raṅkava and Skin

1. Valka (b ark fibres): Kṣuma was one of the most crucial bark fibres in manufacturing 'Kṣauma' cloth. In Rāmāyaṇa (2.6.7), we find the mention of Kṣauma cloth as:

विमलक्षौमसंवीतो वाचयामास स द्विजान् ।

vimalakṣaumasaṁvīto vāchayāmāsa sa dvijān I

We find a mention of kṣauma cloth in Manusmṛti along with (2.41) *saṇa* (flax) and woollen clothes.

शाणक्षौमविकानि च ।

śāṇakṣaumavikāni cha I

Kauṭilya also mentions the presence of kṣauma cloth made in Kāśī and Pauṇḍraka region in his Arthaśāstra

(2.11).

तेन काषिकं पौण्डक्रं च क्षौमं व्याख्यातम् ।

tena kāṣikaṁ pauṇḍakraṁ cha kṣaumaṁ vyākhyātam /

During the Buddhist period in the Pali language, the word 'kṣaumiya' became popular to denote a cloth in general, so the cloth made of cotton was called 'khomiya', which was nothing but the corrupt form of 'kṣaumīya'. This khomiya was used for making 'Civara' of Buddhists Bhikkus.

Kauṭilya, in addition to 'kṣauma', also mentions ūrṇa (wool), valka (bark fibre), karpāsa (cottonseed), tūla (panicle) and atasī (linseed), ṣaṇa (flax), arka, mūltimūrvā, and gavedhuka as fibre yielding plants.

Kauṭilya (2.130), in addition to 'kṣauma' also mention clothes made of wool, bark, cotton, tula and śaṇa.

ऊर्णवल्ककार्पासतूलशणक्षौमाणि

ūrṇāvalkakārpāsatūlaśaṇakṣaumāṇi

From a different reference, we come across that manufacturing of threads (sutra) kots (varma), cloths (vastra) and ropes were supervised by the superintendent of weaving (Kauṭilya, 2.13).

सूत्राध्यक्षः सूत्रवर्मवस्त्ररज्जुव्यवहारं तज्जातपुरूषैः कारयेत् ।

sūtrādhyakṣaḥ   sūtravarmavastrarajjuvyavahāraṁ tajjātapurūṣaiḥ kārayet /

It is also learnt from Arthaśastra (2.11) that the fibres or yarn were extracted from नागवृक्ष, लिकुच, वकुल and वट trees.

नाग वृक्षो लिकुचोवकुलो वटश्च योनयः । पीतिका नागवृक्षका, गोधूमवर्णा लैकुची, श्वेतावकली, शेषा नवीनीत वर्णा ।

nāga vṛkṣo likuchovakulo vaṭaścha yonayaḥ | pītikā nāgavṛkṣakā, godhūmavarṇā laikuchī, śvetāvakalī, śeṣā navīnīta varṇā |

In Addition to cotton, sana (flax) was also cultivated for fibre, according to Manusmṛtri (2.41).

Kauśeya cloths were made of fibres derived from kośa (cocoon), which denotes silk cloth.

In the Vedic period, wool and hair (rāṅkava) were received from the sheep and goat (Ṛgveda 5.59.9). We find the mention of cloth and woollen garments.

उत स्म ते परूष्यामूर्णा वसत शुन्ध्यवः ।

uta sma te parūṣyāmūrṇā vasata śundhyavaḥ.

From the Mahābhārata (2.49.3), we learn that the king of Kambojas presented Yudhisthira woollen clothes of various types, made of sheep's wool, fur, mice, and other animals living in the forest, all in inlaid threads of gold.

In Buddhist sources (*Dīgha Nikāya*, 1.8.40), we also mention blankets made of an owl's feathers. In Āchāraṅga sūtra (2.5.1.6) of Jainas, there is mention of palm leaves as a material for cloths.

वत्थ पडिग्गहं कम्बलं पादपुंछणं उग्गहं च कडासणं एतेसु चेव जाणेज्जा ॥

vattha paḍiggahaṁ kambalaṁ pādapuṁchhaṇaṁ uggahaṁ cha kaḍāsaṇaṁ ētesu cheva jāṇejjā ॥

In Kauṭilya Arthaśāstra (2.11), many wild animals have been mentioned from which furs and skins were obtained for manufacturing woollen clothes. Kauṭilya named their furs and skins '*mṛgaroma*'.

सम्पुटिका चतुरश्रिका लम्बरा वटवानकं प्रावरकः सत्तलिकेति मृगरोमः ।

samputikā chaturaśrikā lambarā vaṭavānakaṁ prāvarakaḥ
sattaliketi mṛgaromaḥ /

Ever since the dawn of civilization, 'wool' as a raw material has occupied an essential place in the textile industry in India.

Textile technology- The progress and advancement of any civilization are assessed from the art of clothing.

It would not be premature to say that spinning, weaving, dyeing, and stitching were in vogue in India when the rest of the world used to live without clothes. According to K.D.V. Codrington, "Since very early times, India has been the significant centre for the allied and linked craft of spinning and weaving" (Parmar, A. Techniques of state craft-Kautilya Arthshartra, Delhi, 1986). According to Chattopadhyaya. (Indian Handicrafts, New Delhi, 1961) "Perhaps India was the first country in the world to have perfected the art and craft of weaving. It is believed that fine weaving went from India to Assyria, Egypt, and through the Phoenicians into Southern Europe.

Techniques of Spinning and Weaving- To manufacture clothes, Indians invented the art of Spinning; before undertaking the processes of spinning, the cotton fibres were cleaned, and the harvested cotton was left in the sun for a few days to weaken to the adherence of seeds to floss in order to facilitate the formers. There were two methods to separate the seeds from floss.

Roller and board

Cotton gin (Charkhi-a machine that separates the seeds, seed hulls, and other small objects from the fibres of cotton)

Both these frames were invented very early. Afterwards, carding (piñjāī) was done to clean, disentangle, and straighten cotton, and then silvers were prepared (pilai). After carding and ginning, spinning was the most essential preliminary process before weaving. The spindle has been the primary instrument for spinning yarn. Once the yarn is spun, it has to be transferred to the 'spool', a wooden winder around which thread can be wound like a bobbin reel; it is known as 'pareti' in Hindi. The collection of yarn in the 'weft spool' to be used in the shuttle was also a critical weaving process.

Here, it may be noted that the spinning and weaving craft developed in the Vedic period itself. In the Ṛgveda (1.142.1; 6.9.2-3), we have a mention of weaving of thread.

तन्तु तनुष्व पूर्ण्यं सुत सोमाय दाशुषे । ऋ. 1.142.1

tantu tanuṣva pūrṇyaṁ suta somāya dāśuṣe I RV. 1.142.1

नाहं तन्तुं न वि जानाभ्योतुं न यं वयन्ति समरे ऽतमाना । ऋ. 6.9.2

nāhaṁ tantuṁ na vi jānābhyotuṁ na yaṁ vayanti samare 'tamānā I RV. 6.9.2

स इत तन्तुं स वि जानात्योतुं स वक्त्वान्यृतुथा ददाति । ऋ. 6.9.3

sa ita tantuṁ sa vi jānātyotuṁ sa vaktvānyṛtuthā dadāti I

RV. 6.9.3

In the *Ṛgveda* (10.130.2), we find the mention of a hand-twisted loom called as 'Tassar'.

इमे मयूखा उप सेदुरे सदः सामानि चक्रु तसराणि ओतवे

ime mayūkhā upa sedure sadaḥ sāmāni chakru tasarāṇi ōtave

In the *Atharvaveda* (10.7.44), we find the same mention.

इमे मयूखा उप तस्तमूः दिवम् सामानि चक्रु तसराणि वातवे

ime mayūkhā upa tastamūḥ divam sāmāni chakru tasarāṇi vātave

The same is still known in India as 'Tussar Silk', also often referred to as 'Wild Silk', Tassar Silk, or Tusar Silk. It is an exquisite thread obtained from a wide-winged moth that is yellowish-brown. The scientific name of these moths is Antheraea Paphia, and they are a part of the group known as Emperor Moths or Saturnids. These moths are a true wonder of nature. Their wings are embellished by circular markings that look like a mirror. Indeed, when you look closely into those circular markings, you'll see a reflection of yourself. Thus, Tussar is highly valued for its natural Gold colours, which it inherits from the Antheria Paphia, Antheria Milita, and Antheria Proylei.

Thus, we know that Tusar silk was known to Indians since the Vedic period. Kosa is the Sanskrit name of Silk or Desi Tussar, produced mainly in India. Koseya, in fact, is the name of Tussar silk in Sanskrit.

Weaver of woollen clothes in the *Ṛgveda* (10.26.2) was called vāyaḥ.

वायः अवीनां आ वासांसि ममृजत्

vāyaḥ avīnāṁ ā vāsāṁsi mamr̥jat

In fact, the English words 'weave' and 'weaver' are the corrupt forms of the Vedic verb √vi and noun vāyaḥ, respectively.

R̥gveda (1.42.1) mentions spinning of thread

तन्तुं तनुष्व पूर्यं सुतऽसोमाय दाशुषे

tantuṁ tanuṣva pūrṇya suta'somāya dāśuṣe

The *R̥gveda* (9.86.32) also refers to a triply twisted thread.

स सूर्यस्य रश्मिभिः परिण्यत तंतुं तन्वानंस्त्रिवृतं यथाविदे ।
नयनृतस्य प्रशिषोनवीयसीः पतिर्जनीनामुपयाति निष्कृतम् ॥

sa sūryasya raśmibhiḥ pariṇyata taṁtuṁ tanvānaṁstrivr̥taṁ yathāvide |
nayanr̥tasya praśiṣonavīyasīḥ patirjanīnāmupayāti niṣkr̥tam ||

Embroidery

Embroidery was another technique of textile designing in ancient India. In the *R̥gveda* (5.74.5; 6.29.3; 10.123.7), we find the mention of 'atka', which was a long and fully covering cloth close to fitting cloak, bright and beautiful, the stuff being bleached cotton interwoven or embroidered cloth with gold threads.

प्रच्यवानात् जुजुरूषः वव्रिम् अत्कम् न मुंचयः । ऋ. 5.74.5

prachyavānāt jujurūṣaḥ vavrim atkam na muṁchayaḥ |
                                        RV. 5.74.5

वसानो अत्कम् सुरभिं दृशे कम्..... ऋ. 6.29.3

vasāno atkam surabhiṁ dr̥śe kam..... RV. 6.29.3

वसानो अल्कं सुरभिं दृशे कं स्वः न नाम जनत प्रियाणि । ऋ. 10.123.7

vasāno atkam surabhim dṛśe kam svaḥ na nāma janata priyāṇi / RV. 10.123.7

उत्शुक्रमत्कमजतं सिमस्मान्नवा मातृभ्यो वसना जहाति । ऋ. 1.95.7

utśukramatkamajatam simasmānnavā mātṛbhyo vasanā jahāti / RV. 1.95.7

In the *Ṛgveda*, we also come across the term 'peśa', which means gold embroidered cloth. It appears that its design was artistic, intricate and inlay of gold heavy and brilliant.

आ एतु पेशा रयिर्भगः स्वस्ति...... । ऋ. 8.31.11

ā ētu peśā rayirbhagaḥ svasti...... / RV. 8.31.11

राजा राष्ट्रानां पेशो नदीनामनुत्तमस्मै क्षत्रं विश्वायुः । ऋ. 7.34.11

rājā rāṣṭrānām peśo nadīnāmanuttamasmai kṣatram viśvāyuḥ / RV. 7.34.11

अधि पेशांसि वपते नृतुरिवापोर्णुतं वक्ष उस्रेव वर्जहम् । ऋ. 1.92.4

adhi peśāṁsi vapate nṛturivāporṇutam vakṣa usreva varjaham. RV. 1.92.4

From the word 'peśa' became the word 'peśaskārī' meaning 'female embroiderer' or 'wife of a maker of gold'.

निस्कृत्यै पेशस्कारीम् । तैब्रा. 3.3.4.5

niskṛtyai peśaskārīm. TBr. 3.3.4.5

The same word later became famous as peśakārī for the art of embroidery. Today, cloth designers are known as peśakāras, and this art is known as peśakārī. In the Vedic literature, the use of words like '*hiraṇyapeśas*' and '*hiraṇyadrāpi*' shows that peśa was a costly cloth inlaid with gold.

Dyeing and washing

Ancient Indians were very well acquainted with techniques of dyeing and washing. Dyeing of textiles was a subsidiary craft. There were used different methods of dyeing clothes. We meet references of several colours in the Vedas in which clothes were usually dyed. The *Ṛgveda* (1.71.1; 1.116-1; 3.44.3; 1.100.16) mentions red, yellow, black, white, green and orange colours.

उप प्र जिन्वन्नुशतीरूशन्तं पतिं न नित्यं जनम सलीलाः ।

upa pra jinvannuśatīrūśantaṁ patiṁ na nityaṁ janama salīlāḥ ।

स्वसारः श्यामीवमरूषीमजूप्रग चित्रमुच्छन्तीमुष्ससं न भावः ॥ ऋ. 1.71.1

svasāraḥ śyāmīvamarūṣīmajūṣraga chitramuchchhantīmussasaṁ na bhāvaḥ ॥ RV. 1.71.1

यमश्विना ददथुः श्वतेनश्रवमघाश्रवाय शश्वदित् स्वस्ति । ऋ. 1.116.1

yamaśvinā dadathuḥ śvatenaśravamaghāśravāya śaśvadit svasti । RV. 1.116.1

अधारयद्हरितोभूरि भोजनं ययोरन्तहरिश्चस्त् । ऋ. 3.44.3

adhārayadharitorbhūri bhojanaṁ yayorantarhariśchast ।
                                          RV. 3.44.3

रोहिच्छयावा सुमदशुर्ललामीर्युक्षा राय ऋप्राश्व । ऋ.1.100.16

rohichchhayāvā sumadaśurlalāmīryukṣā rāya ṛprāśva ।
                                          RV. 1.100.16

Sewing or stitching

The art of stitching dresses has existed in India since the Vedic period. We find references in the Ṛgveda (2.32.4) to sūchi (needle) and the process of sewing.

सीव्यत्वपः सूच्याच्छिद्यमानया ददातु वीरं शतदायमुवश्यम् । 2.32.4

sīvyatvapaḥ sūchyāchchhidyamānayā dadātuṁ vīraṁ
satadāyamuvaśyam I 2.32.4

Various types of clothes mentioned in the ÿgveda also confirm that the art of stitching was invented in India as early as the Vedic period. For instance, the word 'nivi' has been used for the inner garment in the *Atharvaveda* (8.2.16). Above it was worn a cloth 'called 'vasa' and above all was worn 'adhivāsa'. Other names of this 'adhivāsa' were 'atka' and 'drāpi'.

यत् ते वास: परिधानं यां नीविं कृणुषेत्वम् अ. 8.2.16

yat te vāsaḥ paridhānaṁ yāṁ nīviṁ kṛṇuṣetvam a. 8.2.16

अधीवासं परि मातू रिहन्नह तुविग्रेमी। ऋ. 1.140.9

adhīvāsaṁ pari mātū rihannaha tuvigremī I ṛ. 1.140.9

The literature of the later Vedic period and classical Sanskrit is replete with references to various types of clothes. Some of them may be cited as under:

1. Adhivāsa- It was a garment for the upper part of the body.

2. Uṣṇīśa- It was like a turban.

3. Kambala- It was woolen blanket.

4. Drāp- It was a cloak.

5. Vata-pana- It was a wind guard.

6. Samūla- It was like a kurta.

7. Pravara- It was a shawl.

8. Peśas - It was embroidered garment

Organizations engaged in Textile Industry

The textile industry provided occupation to the most

significant number of persons in India, followed by agriculture. So, corporate life was also one of the unique features of ancient India. The organization was the secret to the success of ancient India as it is the secret to the success of the modern developed nations.

One may pose a question who developed this industrial art in ancient India? It is astonishing to know that it was the guild system that was instrumental in the development of industrial art in ancient India. Such a corporate activity is not merely an ordinary feature of ancient Indian life but was the chief secret of its economic success. Guilds are generally considered as associations of merchants or craftsmen for the mutual benefit of their members. They had a monopoly on trade or craft. Even the Vedas indicate the existence of organizations like the guild system, which was nothing but an economic corporation. The word 'Gaṇa' has occurred 46 times in the Ṛgveda and nine times in the Atharvaveda. The term 'Gaṇa' has been defined as assemblies of families.

कुलानां हि समूहस्त गण सम्परिकीर्तितः इति ।

kulānāṁ hi samūhasta gaṇa samparikīrtitaḥ iti /

Thus Gaṇas must have been organised groups for executing some corporate activities. The word 'śreṣṭhī' is quite ancient and occurred in the Vedic literature. For instance, Aitareya Brāhmaṇa (3.30.3) mentions:

....तस्माद् श्रेष्ठी पात्रे रोचयत्येव यं कामयते ...

....tasmād śreṣṭhī pātre rochayatyeva yaṁ kāmayate ...

Taittirīya Brāhamaṇa (3.1.4.10) also mentions the term 'śreṣṭhī' as under:

भगी श्रेष्ठी देवानां स्यामिति। .....श्रेष्ठायाय स्वाहिति।

bhagī śreṣṭhī devānāṁ syāmiti | .....śreṣṭhāyāya svāhiti |

It may be known that guilds were headed by a śreṣṭhi, which later became sethis.

A total of 30 guilds were operating in ancient India, which could be divided into two types.

1. Ordinary guilds.

2. Guilds of technical men or Śilpāyatanas-They included artisans working in wool, silk, cotton, metals, washermen, leathermen, wood, dyers, and dealers in woollen, cotton and silken garment, etc.

Here, it may also be known that guilds were autonomous bodies managed by their members. There was little interference by the state in the affairs of the guild. The people were free to choose their profession according to their interests, merit, and guilds to which they belonged and the area where they wanted to live. In this regard, we get an inkling from Manu (8.41).

जाति जनपदान् धर्माण श्रेणी धर्मश्च धर्मवित्।
समीक्ष्य कुलधर्मांश्च स्वधर्मम् प्रतिपादयेत्।

jāti janapadān dharmāṇa śreṇī dharmaścha dharmavit |
samīkṣaya kuladharmāścha svadharmam pratipādayet |

"A wise man should choose his profession keeping in view the profession of his family, the profession of guilds to which he belongs, the professions in the area (district) he lives in, and his innate nature, interest, and merit.".

Individual Textile Organization

The guilds dealing with the textile industry consisted

of the following individual organisations.

1. Dyer (Vastra-bhedaka): Dyers used to colour the clothes made by weavers. People from the Vedic period itself were familiar with the dyeing technique. Mention of green and red colours in the Ṛgveda (1.14.12)- shows that dyeing was a common practice during that period.

युक्ष्व हि अरुषी रथे हरितः देवः ।
रोहितः ताभिः देवान् इह आ वह ॥

yukṣava hi aruṣī rathe haritaḥ devaḥ /
rohitaḥ tābhiḥ devān iha ā vaha //

Several excavations at Harappa and Mohenjo-daro also confirm the practice of dyeing during that period.

2. **Embroiderer (Peśaskārī):** Embroidery is a very ancient craft in India known even during the Vedic period (Chattopadhyaya et al. Delhi, 1963, p. 43). Indian embroidery was not a simple work of needle and thread. It involved artistic designs and ornamentation. Some of the figurines found at the sites of the Indus Valley Civilization show indications of garments that appear as embroidered. At Mohenjo-daro, the presence of a needle made of bronze confirms that the people of that civilization excelled in embroidery work.

3. **Leatherworkers (Charmakāra):** There are numerous references to leather-made products like upānaha (shoes), dhavitra (air blower of leather), Nṛti (a type of bag), pyukṣā (bow-cover), and abhiṣu (reins of the horse). In addition, we find the presence of leather bags and bellows. All these products prove that some persons were well-versed in cutting, shaping, and sewing leather. They were an essential part of industrial work. Arrian (Indika, p. 219, 230) mentions that Sibai, an Indian tribe, wore

white leather shoes elaborately trimmed while the soles were variegated. Even Strabo (p. 277) and Periplus (39) made several mentions of leather shoes. It shows that Indians were far advanced in leather technology.

4. Tailor (Vstra-rakṣaka): Tailor were known as vastra-rakṣakas in ancient India. As cited above, the references to 'sūchi' in the Vedas prove that the tailor's profession was also prevalent in the Vedic period. In the Rāmāyaṇa (2.80.1; 2.83.2), talented and skilled tailors were called 'Sūtra-karma viśārada'.

अथभूमि प्रदेशज्ञाः सूत्रकर्मविशारदः ।
सूत्रकर्मविशेषज्ञा ये च शस्त्रोप जीविनः ।

athabhūmi pradeśajñāḥ sūtrakarmaviśāradaḥ ।
sūtrakarmaviśeṣajñā ye cha śastropa jīvinaḥ ।

A Bronze needle found in Harappan sites proves that stitching was known then.

In Arthaśāstra (4.1), we come across several references about the tailors who could make any type of garments.

रजकैः तुन्नवायाः व्याख्याता ।

rajakaiḥ tunnavāyāḥ vyākhyātā

Washerman (Rajakas): The word rajaka has appeared in the Yajurveda in the list of various professions, which clearly shows the importance of the profession of rajakas in the Vedic society. Rajakas work came before the dyers. Before dyeing any fabric, it used to be washed by a rajaka. Pāṇini; ग्रामः शिल्पिनी 6.2.62) also mentions the word śilpa in the sense of artisans. Patañjali, in his famous Mahābhāṣya, makes mention of five artisans in each village.

तत्र चावरतः पंचकारुकी

tatra chāvarataḥ paṁchakārukī

5. Weaver (Tantuvāya): As mentioned already, India was the first country on the whole globe to have invented the art and craft of weaving. Indian texts mention many industrial arts, the most prominent being weavers, artisans, etc. The use of separate terms for a weaver, dyer, and embroiderer indicates the workload and specialization in the textile field attained even in the Vedic age, where special mentions are registered of women weavers and (washers) rajakis.

तन्त्रमेके युवती विरूपे अभ्यक्रामं वयतः षण्मुखम् । अ. 10.7.42

tantrameke yuvatī virūpe abhyakrāmaṁ vayataḥ ṣaṇmukham / AV. 10.7.42

Thus, we found the contribution of women in ancient India in all fields, whether it was knowledge proliferation or technology. Women took part in making industrial products on a large scale.

Thus, the textile industry was highly developed in ancient India in all periods till the advent of the British in the 18th century, when they destroyed the entire textile industry of India in order to build their hegemony in the textile field. It may be pointed out here that the textile industry in Lancashire and Manchester was developed on the ruins of the textile industry in India. Textiles were the dominant industry of the Industrial Revolution that began in England in terms of employment, the value of output, and capital invested.

Centres of the Textile Industry in ancient India

India was the home of arts and crafts from time immemorial, and her specialized industries found an appreciative market worldwide. Every house was a hub of some small-scale industrial activity. Cloth was a crucial exchangeable commodity, and its industries in many places were famous for the best quality in the remote past. The following were the main centres of the textile industry in India.

1. **Kalinga (Modern Odisha)**: Nāga people of Odisha earned their name in the art of weaving, which is probably evident from the fact that the word 'Kalinga' in the Tamil language came to denote 'cloth'. Kalingas made the best type of cotton goods. In Mahābhārata, there is a mention that Kalingas, along with Vangas, Tāmarliptas, and Pauṇḍrakas gifted dukula (a kind of excellent cloth which is said to have been made from the fabric obtained from the cotton), Kausika (silk cloth). Patrorṇa (silk and cotton garment), prāvāra (shawl). Arthaśāstra (2.2) mentions Kalinga as one of the seven countries that produced the best type of cotton cloth.

मधुरमापरान्तकं कलिंगकं काशिकं वंगका ..... च कार्पासिकं श्रेष्ठमिति ।

madhuramāparāntakaṁ kaliṁgakaṁ kāśikaṁ vaṁgakā ..... cha kārpāsikaṁ śreṣṭhamiti ।

The Periplus (p. 51-52) states that large quantities of cotton goods, varieties of muslins, and mallow-coloured cotton found their way to the markets of Tagra and Paithan from the country of Maisolos of the Masalia (Kalinga).

2. **Kāmarūpa (Modern Assam)**: Assam has remained famous for its textiles from the very beginning. The fame of Assamese textiles has been recorded even in Harsacharita of Bānabhaṭṭa. The presents sent by King Bhaskarvarman of Assam to Harsha of Kannauj included

a variety of fine textiles. The famous textile product of Assam was linen cloth (Kṣauma) that was bright like the rays of the autumn moon and capable of undergoing washing treatment (śaucha-kṣama).

3. Kāśī (Modern Varanasi): Kāśī cloth was famous for its good quality in the remote past, as mentioned by Kauṭilya (see above). This region has a flourishing textile industry. 'Fine Kāśī cloth worth ten thousand pieces' had become a proverbial phrase. Cotton was cultivated in an extensive field near Banaras. Varanasi also manufactured silk, Kṣauma (linen), and woollen clothes, which were considered very fine.

4. Mathura: Mathura was an essential place from the point of commerce. Mathura was one of the centres of manufacturing clothes. Yuan Chwang quotes that the country of Mathura produced fine striped cotton cloth and gold. (Chakraborty, H- Socio-economic life of India, Cal. 1986, p. 172).

5. Pāṭaliputra: Pāṭliputra was famous for its manufacturing industry in ancient India and Maurayan period. This is confirmed by the fact that a committee of the Municipal Board of Patliputra was especially entrusted with supervising manufactured articles in the metropolis (Dutta et al., An Advance of India, 1982, Delhi, p. 131). There was a separate textile department under the superintendent of weaving called Sutrādhyakṣa. One may safely say that to ensure uninterrupted textile production and distribution,

Pāṭaliputra played a major role by providing a controlled market.

6. Ujjain: Ujjain, known by Malva and Awanti, was

another textile hub of ancient India, as is evident from Periplus. India's wealth was enhanced by the cotton industry, which rose to its highest stage of advancement. Not only Ujjain in the west, Malwa was a centre of the cotton weaving industry because of its fertile black soil. This is evident from an epigraphic record of Sanchi, which records the gift of the weaver (Sotika) and indicates that weavers had concentrated themselves in and around Vidisa. The reference to a village called Kapasigram, near Vidisa, noted for cotton and cotton fabrics, is insignificant. Ujjaini and Vidisa also occupied positions of commercial importance due to their location on the crossing trade routes.

7. **Vaiśālī**: It was the capital of Lichchhavis and the headquarters of the Ujjaini confederacy. Vaiśālī was a great market town in Eastern India and the chief market town of North Bihar. The textile industry of this land was highly developed, as we find a mention in the Rāmāyaṇa (Balkanda, 74 Sarga) that King Janaka gave numerous blankets, silk, or linen garments to Sita in her marriage.

अथ राजा विदेहानां ददौ कन्याधनं बहु ।
गवां शतसहस्राणि बहूनि मिथिलेश्वरः ।
कम्बलानां च मुख्यानां क्षौम कोट्यम्बराणि च ॥

atha rājā videhānaṁ dadau kanyādhanaṁ bahu /
gavāṁ śatasahasrāṇi bahūni mithileṣvaraḥ /
kambalānāṁ cha mukhyānāṁ kṣauma koṭyambarāṇi
cha //

8. **Vaṅga (Modern West Bengal)**: In the hoary past of Indian civilization, Bengal was at its peak in prosperity as an agricultural, commercial, and manufacturing hub. Though agriculture formed the most predominant

feature of Bengal's economy, various arts and crafts were also developed over time (Majumdar, 1978, p. 341). As a trading region, Bengal was known to the ancient people bordering the Mediterranean Sea. It will be seen from the fact that its muslin fabrics were the most delicate and extremely beautiful. They were in great demand under the name of 'Gangetika', which suggests their origin close to the banks of the Ganges.

Bengal attained a great name in the textile industry in the ancient period, confirmed by the Arthaśāstra of Kauṭilya (2.11), which includes the following best qualities of the fabrics of various kinds and soft like a gem and worked in gold.

1. Dukula (white and soft fabric) manufactured in Vaṅga called Vaṅgaka.

2. Dukula (black and soft like the surface of a gem) manufactured in Pauṇḍra kingdom (modern-day Bangladeśa and West Bengal)

3. Kṣauma (muslin) manufactured in Pauṇḍra kingdom (present-day Bengladeśa and West Bengal).

4. Patrorna or Pauṇḍra

5. Kapāsika (cotton fabrics)

वांगकं श्वेतं स्निग्धं दुकूलं पौण्ड्रकं श्यामं मणिस्निग्धंए, सौवर्णकुड्यकं, सूर्यवर्ण मणिस्निग्धौदकवानं चतुरश्रवानं व्यमिश्रवाणं च।

vāṁgakaṁ śvetaṁ snigdhaṁ dukūlaṁ pauṇḍrakaṁ śyāmaṁ maṇisnigdhamē, sauvarṇakuḍyakaṁ, sūryavarṇa maṇisnigdhaudakavānaṁ chaturaśravānaṁ vyamiśravāṇaṁ cha /

Gujrat: The cotton cloth was manufactured in all parts of Northern and Western India. Undoubtedly,

Gujrata was one of the most critical seats of production. Ev en Mānasollāsa (3rd century AD) gives us a long list of fabrics and places where they were made. He says that Gujrat produced for export to Arabian lands a high quality of foreign cotton of every colour. Marco-polo, a traveller of the 15th century, observes that much fine buckram (cotton stuff) was produced in the kingdom of Cambay (Khambat) and Mālābār (India's southwestern coast, which lies on the narrow coastal plain of Karnataka and Kerala states between the Western Ghats range and the Arabian Sea. The coast runs south of Goa to Kanyakumari on India's southern tip). He also mentions that they were exported from Gujrat to other places. Thus, we see that in the ancient period, Gujrat was a hub of the textile industry, and today, it is a leading state of India in textile manufacturing.

Aprānta (Northern part of the Konkan region on the western coast of India): The Arthaśāstra of Kauṭilya (2.2) gives us to understand that Aparānta was a cotton-growing region in ancient India. Māhābhārata also suggests to us that Aparānta was a cotton-producing country. The people of Aprānta presented the Paṇḍavas gift of smooth white cotton garments. (Das, D.R. Economic History of Deccan, Cal. 1969, p.98) Even today, *regar* black soil in the valley of Tapti, Godavari, Narmada, Krishna, and Northwest of Maharastra, is the main cotton-growing soil in India. Divyavadāna refers to a class of cotton called Aparāntaka, which suggests its manufacturing in Aparāta. The exact text speaks of another variety of textiles called 'Phuttaka'. It was probably a variety of printed calico cloth. Kauṭilya (2.2) adjudges the cotton manufactured in Aparānta was of high quality.

मधुरमापरान्तिकं ....कार्पासिकं श्रेष्ठम् ।

madhuramāparāntikaṁ ....kārpāsikaṁ
śreṣṭham I

It was exported from Bhroach to the West.

Textiles Centres of Dakṣiṇāpatha: South India has remained a significant hub of cotton and silk clothes from the ancient period onward. We learn from the Mahābhārata that the present starting form of clothes made of wool of sheep and goats, even muslin, was given to King Yudhiṣṭhira by the people of Karnataka and Mysore. Evidence of what spindles were found in archaeological sites like Paiyampalli (in Tirupattur Talauq) in North Arcot district, Adichchanallur, and Nilgiri hills of Tamilnadu. Warangal was famous for manufacturing carpets, which were in great demand. Morcopolo, a traveller of the 15th century A.D., gives his views about the quality of the products of the Warangal region:

These are the most delicate buckrams (cotton stuff) and of the highest price; in the south, they look like the tissue of a spider's web. There can be no king or queen in the world who might not be glad to wear them.

The above observation of Marcopolo proves that the fabric of Tamilnadu achieved a great degree of technical skill and enjoyed a great deal of popularity in ancient periods.

Kauṭilya (2.2) appreciates the cotton manufactured in Mahisa (Maharaṣṭra) and the Vats region (Vatapi or Badami region in the Bagalkot region of Karnataka) as the best one.

वात्सकं महिसकं च कार्पासिकं श्रेष्ठम् ।

vātsakaṁ mahiṣakaṁ cha kārpāsikaṁ śreṣṭham /

Periplus (section 60) refers to Poduca (Pondicherry), whose historical importance as a textile hub is evident from Arikkamedu's excavations. (Ramaswamy, V. Textiles and Weavers in Medieval South India, Madra, 1985, p.3). Huge brick dyeing vats from the first and second centuries have been unearthed from Arikkamedu, in Pondicherry and Uraiyur in Tiruchirapalli, both known as the critical weaving centre from the accounts of the Periplus (Ramasvami, 1985, p. 2).

Export: Thus, we have seen from the discussion above that Indians were producers of high-quality cotton, silk, muslin, and wool. They also developed advanced technology to manufacture textile products. Around 700 B.C., the cotton tree was introduced in Assyria from India during the reign of Sennacherib. Indians were also the exporters of raw materials and finished products of textiles to the world's various countries. The kings and queens of various countries were fond of wearing Indian clothes. According to C.V. Vaidya (1933: 236), 'Gold latticed silken, cotton and woollen cloth used to be exported from India by land and sea routes in the early times.'

Warmington also referred to the export of raw cotton for manufacture abroad, saying that "from 100 AD onwards, much raw cotton was submitted to the looms of Alexandria and Syria." (Sastri,1966, p. 337)

Stein (1907: pp. 374,412,422 & 460) refers to cotton cloth cultivation and foreign export. Based on these records, it can be unhesitatingly maintained that cotton fabric spread from India to Central Europe as early as the 1st century B.C.

14

Instrumentation

Various apparatus and their applications in the Ayurveda

Āchārya Somadeva took a leading role in describing various instruments (apparatus) used in Ayurveda. Rasaratnasamucchaya also quotes him in the ninth chapter while dealing with various Ayurvedic instruments. Rasaratsamucchaya defines the Ayurvedic Instrument as under.

स्वेदादिकर्मनिर्मातुं वार्तिकेन्द्रैः प्रयत्नतः ।
यन्त्र्यते पारदो यस्मात्तस्माद्यन्त्रमिति स्मृतम् ॥

svedādikarmanirmātuṁ vārtikendraiḥ prayatnataḥ /
yantrayate pārado yasmāttasmādyantramiti smṛtam //

[Meaning] Yantra is an instrument used to have a controlled application of mercury for sweating, etc., actions of Ayurveda.

In Rasaratnasamucchaya (Chapter 9), various apparatus (instruments) have been described. Some of them are described hereunder:

1. *Dolāyantra*: A pot is half-filled with a liquid and a rod placed across its mouth from which the medicine is suspended and tied in a piece of cloth. The liquid is allowed to boil, and a second pot is inverted over the first.

द्रवद्रव्येण भाण्डस्य पूरितार्धोदकस्य च ।
मुखस्योभयतो द्वारद्वयं कृत्वा प्रयत्नतः ॥ 9.3

dravadravyeṇa bhāṇḍasya pūritārdhodakasya cha |
mukhasyobhayato dvāradvayaṁ kṛtvā prayatnataḥ || 9.3

तयोस्तु निक्षिपेद्दण्डं तन्मध्ये रसपोटलीम् ।
बद्धा तु स्वेदयेदेतद्दोलायन्त्रमिति स्मृतम् ॥ 9.4

tayostu nikṣipeddaṇḍaṁ tanmadhye rasapoṭalīm |
baddhvā tu svedayedetaddolāyantramiti smṛtam || 9.4

2. A pot with boiling water has its mouth covered with a piece of cloth, and the substance to be steamed is placed on it, and a second pot is arranged in an inverted position over the rim of the first. (*Rasaratnasamucchaya*, 9.4)

3. *Pātanā yantra*: (Apparatus for sublimation and distillation): Two vessels are adjusted so that the neck of one fits into that of the other. The junction of the necks is luted with a composition made of lime, raw sugar, rust of iron, and buffalo milk (*Rasaratnasamucchaya*, 9.6-7).

4. *Adhaspātanā yantra*: A modification of the PÈtanÈ apparatus in which the bottom of the upper vessel is smeared with the substance, the vapour or essence thereof condensing into the water of the upper vessel using the fire of dried cow-dung cakes. (*Rasaratnasamucchaya*, 9.9)

5. *Ḍekī yantra*: Below the neck of the pot is a hole into which the upper end of a bamboo tube is introduced, the lower end of it fitting into a brass vessel filled with water and made of two hemispherical halves. Mercury mixed with the proper ingredients is distilled until the receiver gets sufficiently heated. (*Rasaratnasamucchaya*, 9.17-19)

6. *Bāluka yantra*: (Sand bath apparatus): A glass with a long neck containing mercurials is wrapped with several folds of cloth smeared with clay and then dried in the sun. The flask is buried up to three-fourths of its lengths in the sand and placed in an earthen pot whilst another pot is inverted over it, the rims of both being luted with clay. Heat is now applied till a straw placed on its top gets burnt. (*Rasaratnasamucchaya*, 9.36-39)

7. *Lavaṇa yantra*: (salt bath): If in the Bāluka yantra salt is substituted for sand, it is called lavaṇa yantra. (*Rasaratnasamucchaya*, 9.40)

8. *Nālikā yantra*: If in the Bāluka yantra, an iron tube is substituted for the glass flask, it is called Nālika yantra. (*Rasaratnasamucchaya*, 9.43)

9. *Bhūdhara yantra*: Place the crucible containing chemicals inside a sand mass and apply heat through cow-dung cakes. This is known as BhÊdhara yantra. (*Rasaratnasamucchaya*, 9.44)

10. *Tiryakpātana yantra* (**distillation**): Place the chemicals in a vessel with a long tube inserted in an inclined position, which enters the interior of another vessel arranged as the receiver. The mouths of the vessels and the joints should be luted with clay. Now urge an intense fire at the bottom of the vessels containing the chemicals, whilst in the other vessel place cold water. This process is known as Tiryakpātana. (*Rasaratnasamucchaya*, 9.10-12)

11. *Vidyādhara yantra*: Vidyādhara yantra is for the extraction of mercury from hiṅgula (cinnabar). Two earthen pots are arranged so that the upper one contains cold water, and mercury condenses at its bottom

(*Rasaratnasamucchaya*, 9.56-57).

12. ***Dhūpa yantra*** (Fumigating apparatus): Bars of iron are laid in a slanting position a little below the mouth of the lower vessel, and gold leaves are placed over them and at the bottom of the vessel is deposited a mixture of sulphur, realgar orpiment, etc. A second vessel, with its convexity turned upwards, covers the mouth of the lower one and the brims are luted with clay. Heat is now applied from below. This is called the fumigation of the gold leaves. Silver may also be similarly treated (*Rasaratnasamucchaya*, 9.69-73).

An Apparatus for Killing Metals:

According to Rasārṇava ((Ray, 1903: p.65), two iron crucibles, each of 12 digits in length, with a narrow orifice containing sulphur, is inserted into the other holding mercury; below the mercury is placed water [in a separate cvessle]. The mercury and sulphur should be carefully moistened in garlic juice filtered through a cloth. The apparatus is now lodged in an earthen pot and another placed over it, the rims being lited with a cloth previously smeared with earth; now, the cow-dung fire is urged. After continuing heating for three days, it resembles the nipple of a cow and is fitted with a lid, which has a raised head.

For the purification of silver, the crucible is best made of two parts of the ashes of Schrebera swietenioides and one part each of the brick dust and earth.

Various apparatus and their applications in metallurgy

Various instruments/apparatus, viz. Muṣā Koṣṭhi

Yantra, etc., and their applications for the extraction of various metals in pure form are available from the ancient Sanskrit texts dealing with metals and metallurgy.

For instance, *Rasārṇava* and *Rasaratna Samucchaya* describe a typical crucible, known as *vrintak*, with a shape similar to that of an extended variety of brinjal used for making the reduction-distillation chamber. The basic principle of the process resembles that of the large-scale 12-century industrial process for zinc extraction uncovered at Zawar near Udaipur. It is a unique discovery, and the retorts used at Zawar are similar to the *Vrintak* crucible.

Surgical Instruments

The surgery was also in practice in ancient India. The Ayurvedic texts have dealt with the surgery of various kinds and surgical instruments. Surgical instruments have been classified into the yantras and the śastras, i.e., the blunt and the sharp instruments. Suśruta enumerates no less than one hundred varieties of blunt instruments and twenty kinds of sharp instruments. Hārīta, on the other hand, enumerates twelve blunt instruments, twelve sharp instruments, and four prabandhas as necessary for the operation of the extraction of arrows and other foreign bodies.

द्वादशेव तु यन्त्राणि शास्त्राणिद्वादशैव तु ।
चत्वारि च प्रबन्धानां शल्योद्धारे विनिर्दिशेत् ॥

dvādaśeva tu yantrāṇi śāstrāṇidvādaśaiva tu ǀ
chatvāri cha prabandhānāṁ śalyoddhāre vinirddiśet ǀǀ

Vāgbhaṭa II mentions one hundred and fifteen kinds

of blunt and twenty-six kinds of sharp instruments. Pālakapya (Treatment of Elephants) mentions ten kinds of śastras or sharp instruments, though he describes the uses of other instruments required for the surgical treatment of diseases.

Blunt Instruments and their classification

Of the one hundred and one Varieties of the blunt instruments, the surgeon's hand is rightly considered as the principal instrument for without its help, no instrument can properly be used, and every surgical operation is under its control.

यन्त्रशतमेकोत्तरमत्र हस्तमेव प्रधानतमं यन्त्राणामवगच्छ। किं कारणं। यस्माद्धप्तादृशे यन्त्राणामप्रवृत्तिरेव तदधीनत्वाद्यन्त्रकर्म्मणां।

तत्र मनःशरीरवाधकराणि शाल्यानि तेषामाहरणी पायो यन्त्राणि।

तानि षड्प्रकाराणि। तद्यथा। स्वस्तिकयन्त्राणि। सन्दंशयन्त्राणि। तालयन्त्राणि। नाडीयन्त्राणि। शलाकायन्त्राणि। उपयन्त्राणिचेति। तत्र चतुर्व्विंशति स्वस्तिकयन्त्राणि। द्वे सन्दंशयन्त्रे। द्वे एव तालयन्त्रे विंशतिर्नाड्यः। अष्टविंशतिः शलाकाः। पंचविंशति रूपयन्त्राणि।

yantraśatamekottaramatra   hastameva   pradhānatamaṁ yantrāṇāmavagachchha ǀ kiṁ kāraṇaṁ ǀ yasmāddhaptādṛśe yantrāṇāmapravṛttireva tadadhīnatvādyantrakarmmaṇām ǀ

tatra manaḥśarīravādhakarāṇi śālyāni teṣāmāharaṇī pāyo yantrāṇi ǀ

tāni   ṣaṭprakārāṇi ǀ   tadyathā ǀ   svastikayantrāṇi ǀ sandaṁśayantrāṇi ǀ   tālayantrāṇi ǀ   nāḍīyantrāṇi ǀ śalākāyantrāṇi ǀ upayantrāṇicheti ǀ tatra chaturvviṁśati svastikayantrāṇi ǀ dve sandaṁśayantre ǀ dve ēva tālayantre viṁśatirnāḍyaḥ ǀ   aṣṭaviṁśatiḥ   śalākāḥ ǀ   paṁchaviṁśati rūpayantrāṇi ǀ

They are recommended to be used for the extraction

of śalya or foreign bodies, e.g., a dar, an arrow, a javelin, a spear, a peg, a pin, a bamboo rod, a stake etc., which cause pain to the body and mind.

A. Suśruta sub-divides the blunt instruments into six classes, viz.:

i.	Svastika or cruciform instruments	24 kinds
ii.	Sandañśa or pincher-like	2 " "
iii.	Tāla or picklock-like	2 " "
iv.	Nāḍī or tubular of hollow	20 " "
v.	Śalākā or rod or pricker-like	28 " "
vi.	Upayantra or accessory	<u>25 " "</u>
		101" "

These instruments are advised to be made generally of iron or other suitable materials when iron is unavailable. Their ends often resemble the faces of some ferocious beasts, deer, or birds. Hence, the instruments should be so constructed as to have the likeness of their faces, following at the same time the directions of the scientific treatises, the instructions of the teachers or imitation of other instruments, or in adaptation to the exigencies of the time. They should be reasonably sized, with their ends rough or smooth as required. They should be of solid make, good shape, and capable of a firm grasp.

तानि प्रायशो लौहानि भवन्ति तत्प्रतिरूपकाणि वा तदलाभे। तत्र नानाप्रकाराणां व्यलानां मृगपक्षिणां मुखैर्मुखानि यन्त्राणां प्रायशः सदृशानि तस्मात्तत्सारूप्यादागमादुपदेशादन्ययन्त्रदर्शनाद्युक्तितश्च कारयेत्। समाहितानि यन्त्राणि खरश्लक्ष्णमुखानि च।

सुदृढानि सुरूपाणि सुग्रहाणि च कारयेत्॥ Suśruta Saṅghtā, I. vii.

tāni prāyaso lauhāni bhavanti tatpratirūpakāṇi vā

tadalābhe I tatra nānāprakārāṇāṁ vyalānāṁ mṛgapakṣiṇāṁ mukhairmukhāni yantrāṇāṁ prāyaśaḥ sadṛśāni tasmāttatsārūpyādāgamādupadeśādanyayantradarśanādyuktitā ścha kārayet I samāhitāni yantrāṇi kharaślakṣaṇamukhāni cha I

sudṛḍhāni surūpāṇi sugrahāṇi cha kārayet II

अनेकरूपकार्य्याणि यन्त्राणि विविधान्यतः ।
विकल्प्य कल्पयेद्रॄ बुद्ध्या यथास्थूलन्तु वक्ष्यते ॥
अलोहान्युशस्त्रातान्येव च्व विकल्पयेत् ।
अपराण्यापि यन्त्रादौन्यपयोगंच यौगिकम् ॥ *Aṣṭāṅga Hṛdaya, I. xxv.*

anekarūpakāryyāṇiā yantrāṇi vividhānyataḥ I
vikalpya kalpayeḍṛ buddhyā yathāsthūlantu vakṣayate II
alohānyuśasrātānyevachcha vikalpayet I
aparāṇyāpi yantrādaunyapayogaṁcha yaugikam II

I.   The Swastika or Cruciform instruments are –

1. Siṅgamukha 2. Vyāghramukha 3. Vṛkamukha

4. Tarakṣumukha 5. Ṛkṣamukha 6. Dvīpimukha

7. Mārjāramukha 8. Śṛgālamukha

9. Airvvārukamukha 10. Kākamukha

11. Kaṅkamukha 12. Kuraramukha 13. Chāsamukha

14. Bhāsamukha 15. Śaśaghātīmukha

16. Ulūkamukha 17. Chillimukha 18. Gṛdhramukha

19. Śyenamukha 20. Krauñcamukha

21. Bhṛṅgarājamukha 22. Añjalikarṇamukha

23. Avabhañjanamukha and

24. Nandimukhāmukha

II. The Sandañśa or pincher-like instruments are–

1. Forceps with arms

2. Forceps without arms

III. The Tāla or picklock-like instruments are –

1. Ekatāla

2 Dvitāla

IV. The Nāḍī or tubular instruments are –

For Fistula-in-ano– (1) with one slit; (2) with 2 slits	2
For piles – (1) with one slit; (2) with 2 slits	1
For Wound	1
For Clysters (Rectal)– (Some authors describe 3 only)	4
For clysters (Vaginal and urethral)– (male and female)	3
For Hydrocele	1
For Ascites	1
Fumigation and inhalation	3
For Urethral Stricture	1
For Rectal	1
For Cupping gourd	1

	20

V. The Śalākā or rod-shaped instruments are –

Gaṇḍūpadarmukha or earth-worm like	2
Śarapuṅkhamukha or arowl-stem like	2
Sarpafaṇamukha or snake's hook like	2
Vaḍiśamukha or fish-hook like	2

Masūradalamukha or masūra pulse like	2
Promārjana or swabs	6
Khailamukha or spoons	3
Jāmvavavadana or jambul seed like	3
Ankuśavadana or goad like	3
Kolastinidalamukha or plum seed like	1
Mukulāgra or bud shaped	1
Māltipuṣpavṛntāgra or like stem of mālati flower	1
	28

VI. The Upayantra or accessory instruments are –

1. Rajju (thread)

2. Veṇikā Twine)

3. Paṭṭa (bandages)

4. Charma (leather)

5. Valkala (bark of trees)

6. Lats (creepars)

7. Vastra (cloth)

8. Aṣṭhīlāśma (stone or pebble

9. Mudgara (hammer)

10. Pāṇipādatala (palm of the hand or sole of the foot).

11. Aṅguli (finger)

12. Jihvā (tongue)

13. Danta (tooth)

14. Nakha (nail)

15. Mukha (mouth)

16. Vāla (hair)

17. Aśvakaṭaka (the ring of a horse's bridle)

18. Śākhā (branch of a tree)

19. Sṭhīvana (spittle)

20.Pravāhana (flushing the patient)

21. Harṣa (objects exiting happiness)

22. Ayaṣkānta (a loadstone)

23. Kṣāra (caustic)

24. Agni (fire)

25. Bheṣaja (medicines).

B. The sharp instruments or śastras are –

1.   Maṇḍalāgra or roundheaded knife

2.   Karapatra or saw (lit like the human hand)

3.   Vṛddhipatra (lit. like the leaf of vṛddhi–an unknown medicinal plant)–a razor

4.   Nakha-śastra or nail-parer

5.   Mudrikā or finger-knife (like the last phalanx of the index finger)

6.   Utpalapatra, a knife, i.e. resembling the petal of blue lotus

7.   Arddhadhāra or a single-edged knife

8.   Sūcī–needles

9.   Kuśapatra–a knife shaped like the kuśa grass

10. Ātīmukha–a knife like the beak of the Ātī bird (Turdus Ginginianus)

11. Śararī-mukha–a "pair of scissors like the breaks of Śararī bird

12. Antarmukha (lit. having internal sharp edge)– a kind of scissors

13. Trikurccaka– an instrument consisting of three needles

14. Kuṭhārikā–a small axe shaped instrument

15. Vrīhmukha–a trocar shaped like a grain of rice

16. Arā or awl

17. Vetasa-partaka–an instrument shaped like the leaf of a rattan (Calamus Rotang)

18. Vaḍiśa–an instrument shaped like the fish-hook

19. Dantaśaṅku or tooth-pick

20. Eṣani or sharp probe-like instruments

According to Hārīta, the twelve blunt instruments are-

1. Godhāmukha or iguna-faced 2. Vajramukha-gṛdhramukha. 3. Trivaktra or three faced. 4. Sandañśa or pincher. 5. Cakrākṛti or circular shaped. 6. Anaka. 7. Kaṅkapāda. 8. Śṛṅga or horn. 9. Kuṇḍala. 10. Śrīvatsa. 11. Sauvatsika. 12. pañcavaktram, i.e., five faced-siṅhamukham.

गोधामुखं वज्रमुखं त्रिवक्रं नाम सन्दंशरचक्रकृतिकङ्कपादम् ।
अथानकं श्रृङ्गककुण्डलं च श्रीवत्ससौवत्सिक पंचवक्रं ।
द्वादशैतानि यन्त्राणि कथतानि भिषम्वरैः ।
अथ शस्त्राणि प्रोक्तानि नामानि च पृथक् पृथक् ।

अर्द्धचन्द्रं त्रीहिमुखम् कङ्कपत्रं कुटारिका ।
करवीरकपत्रं च शलाककरपत्रकम् ।
वड्शिं गृध्रपादं च शूली च सूचिमुद्ररम् ।
शस्त्रधयेतानि प्रोक्तानि शल्येद्धारे पृथक् पृथक् ॥

Hārīta Saṁhitā. 3.6

*godhāmukhaṁ vajramukhaṁ trivaktraṁ nāma
sandaṁśarachakrakṛtikaṅkapādam ǀ
athānakaṁ śṛṅgakakuṇḍalaṁ cha śrīvatsasauvatsika
paṁchavaktraṁ ǀ
dvādaśaitāni yantrāṇi kathatāni bhiṣamvaraiḥ ǀ
atha śastrarāṇi proktāni nāmāni cha pṛthak pṛthak ǀ
arddhachandraṁ vrīhimukham kaṅkapatraṁ kuṭārikā ǀ
karavīrakapatraṁ cha śalākakarapatrakam ǀ
vaḍśiṁ gṛdhrapādaṁ cha śūlī cha sūchimudgaram ǀ
śastradhayetāni proktāni śalyeddhāre pṛthak pṛthak ǁ*

The Twelve Sharp Instruments of Hārāta are –

1. Arddhacandra or half-moon shaped. 2. Vrīhimukha. 3. Kaṅkapatra. 4. Kuṭhārikā. 5. Karavīrkapatraka. 6. Śalākā or sharp probe. 7. Karapatraka or saw. 8. Vaḍiśa or sharp hook. 9. Gṛdhrapāda. 10. Śulī or needle. 12. Mudgara or hammer.

Vāgbhaṭa II classifies the instruments in the following way–

A. Blunt instruments –

Sr. No.	Name	No.
I.	Svastika, as heron, lion, bear, crow, deer, forceps, etc.	24
2.	Sandañśa: It consists of two iron blades soldered at one end, the other ends being free	2

	(a) for extraction of eyelashes etc.	1
	(b) muchuti	1
3.	**Tāla**	**2**
	(a) Ekatāla	1
	(b) Dvītāla	1
4.	**Nāḍi or tubular**	**23**
	(a) Kaṇṭhaśalyāvlokinī or throat speculum having free and five holes	2
	(b) Śalyanirghātanī	1
	(c) For piles, different sizes for male and female For inspection: two holes-rectal speculum For medication: 1 hole For applying pressure: entire-śāmī	6 2 2 2
	(d) For fistula-in-ano : with one-two holes	2
	(e) For nasal polypus etc.	1
	(f) Aṅguli-trāṇaka or finger-gaurd	1
	(g) joni-vraṇekṣaṇa or vaginal speculum	1
	(h) Vrāṇo vasti or wound syrmge	1
	(i) For dakodara or Parenthesis abdominis	1
	(j) Vastiyantra or clysters: rectal, vaginal, and urethral	3
	(k) For fumigation	1
	(l) Cupping instruments: Alābu, Ghaṭīyantra	3

	and Horns	<u>23</u>
5.	**Śalālā or rod-like instruments**	**34**
	(a) Gaṇḍūpadamukha or earth-worm shaped	2
	(b) Masūradalabaktra	2
	(c) Śaṅku	9
	Faṇībaktra or snake's hood	2
	Śarapuṅkamukha	2
	Vaḍiśa or blunt hook	2
	Garvaśaṅku or delivery hook	1
	Aaśmarī or lithotomic hook	1
	Śarapuṅkamukha or tooth extractor	1
	(d) For wiping out discharges	6
	For rectum	2
	For nose	2
	For ears (karṇaśodhana)	2
	(e) For the application of actual and potential cauterise	11
	Jāmvobou——ha, three for each	6
	Arddhendu or half-moon shaped, for hernia	1
	Kolāsthidala for nasal polypus	1
	Nail-shaped	3
	(f) For cleansing	1

	Rectum	1
	Vagina	1
	Urethra	1
	(g) Collyrium robe	<u>1</u>
		<u>34</u>

VI. Anuyantra or accessory instruments are nineteen in number. To the list of Suśruta he (*Aṣṭāṅga Hṛdaya*, 1-25) adds the following –

Goat's gut, silk, time, suppuration, and fear.

अनुयन्त्राण्यस्कान्त रज्जु वस्त्राऽश्म मुद्गराः ।
पट्टान्त्र जिह्वा वालाश्च शाखा नख मुख द्विजाः ॥
कालः पाकः करः पाद्मोभयं हर्षश्च तत् क्रियाः ।
उपायवित् प्रविभजेदालोच्य निपुणं धिया ॥ ।

anuyantrāṇyaskānta rajju vastrā'śma mudgarāḥ |
paṭṭāntra jihvā vālāścha śākhā nakha mukha dvijāḥ ||
kālaḥ pākaḥ karaḥ pādmobhayaṁ harṣaścha tat kriyāḥ |
upāyavit pravibhajedālochya nipuṇaṁ dhiyā || |

B. The sharp instruments of Vāgbhaṭa are twenty -six in number–

1. Maṇḍalāgra 2. Vṛddhipatra 3. Utpalapatra

4. Adhyarddhadhāra 5. Sarpāsya

6. Eṣaṇḍūpadamukhā 7. Vetasa 8. Śarārī

9. Trikurccaka 10. Kuśapatra 11. Ātīvadana

12. Antarmukha and Arddhacandrānan

13. Vrīhibaktra 14. Kuṭhāri 15. Kura-vakasalā

16. Aṅglīśastra 17. Vaḍiśa 18. Karapatra 19. Kartarī

20. Nakhaśastra 21. Dantalekhana 22. Sūcī

23. Kurcca 24. Khaja 25. Ācā 26. Karṇavedhanī

षड्विंशतिः सुकमारैर्घटितानि यथाविधि ।
शस्त्राणि रोमवाहीनि वाहुल्येनाङ्गलानि षट् ॥
सुरूपाणि सुधाराणि सुग्रहाणी च कारयेत् ।
अकरलानि सुध्मात सुतीक्ष्णावार्त्तऽयसि ॥

ṣaḍviṁśatiḥ sukamārairghaṭitāni yathāvidhi |
śastrāṇi romavāhīni vāhulyenāṅgalāni ṣaṭ ||
surūpāṇi sudhārāṇi sugrahāṇī cha kārayet |
akaralāni sudhmāta sutīkṣaṇāvārtta'yasi ||

Bhāvamiśra mentions the following blunt and sharp instruments: Eṣanī, Jāmvousṭha Śalā, Sūcī and knives generally in making incisions which should be shaped like Kharjjūrapatrika, (like the leaf of Kharjjūra tree, Phoenix Sylvester's. Roxb) Arddhacandra, Candravarga, Sūcimukha, and Abāṅmukha.

एषण्या गतिमन्विष्य श्वारसूत्रानुसारिनीम् ।
सूचीं निद्ध्यादत्यन्ते प्रोत्राम्याशु विनिदरेत् ॥
(Bhāva Prakāśa, 2.4, Nāḍivraṇādhikāra)

ēṣaṇyā gatimanviṣya ścharasūtrānusārinīm |
sūchīṁ niddhyādatyante protrāmyāśu vinidaret ||
(Bhāva Prakāśa, 2.4, Nāḍivraṇādhikāra)

अगन्तुजं भिषग्ग्राड़ी शास्त्रेणोत्कृत्य यत्नतः ।
जाम्बोष्ठोनाग्निवर्णेन तस्या वा शलाकया ॥
(Bhāva Prakāśa, Bhagandharādhikāra)

agantujaṁ bhiṣagnārī śāstreṇotkṛtya yatnataḥ |
jāmboṣṭhonāgnivarṇena taptyā vā śalākayā ||
(Bhāva Prakāśa, Bhagandharādhikāra)

गतिमन्विष्य शस्त्रेण दिन्दरात् खर्ज्जूरपत्रिकम् ।
चन्द्रार्द्धे चन्द्रवर्गंच सूचीमुखमवाङ्खुखम् ॥ (*ibid.*)

gatimanviṣya śastreṇa dindarāt kharjjūrapatrikam /

chandrārddha chandravargaṁcha

sūchīmukhamavāṅmukham // (ibid.)

Pālakāpya (Hastī Āyurvea, 3.30) mentions ten kinds of śastras:– 1. Vṛddhipatra 2. Kuśapatra. 3. Maṇḍalāgra. 4. Vrīhimukha. 5. Kuṭhāri. 6. Vatsadanta. 7. Utpalapatra. 8. Śalākā. 9. Śūcī or needles. 10. Rampaka. Besides these he refers to Vaḍiśa.

Of the blunt instruments, he mentions Jāmvobouha- (for application of actual cauterise), Siṁhadaṁṣṭrā, Godhāmukha, Kaṅkamukha, Kuliśamukha (for extraction of foreign bodies), Eṣaṅī or probe (three), wound syringe, Vastiyantra, Śalākā or rods, Yaṣṭhiyantra, Karkataka, Dyātūha, Makaraka (crocodile), Śārddūlamuṣṭhika (tiger's claws), Nandimukha (Turdus ginginianlus).

तत्र शस्त्राणि दशनामसंस्थानानि भवन्ति। तद्यथा वृद्धिपत्रम्, कुशपत्रम्, मण्डलायम् व्रीहिमुखम्, कुठराकृति, वत्सदन्तम्, उत्पलपत्रमए शलाकाए सूचीए रम्पकश्चेति। फालजामववतापिकार्य्यकृत्यश्चेति। एतान्धाग्निकर्म्मविधाने चत्वारि चान्यानि शलयद्धरणानि यथोयोगं सिंहद्रष्टं गोधमुखं कङ्कमुखं कुलिशमुखं चेति। तिस्र एषिणयः।

tatra śastrāṇi daśanāmasaṁsthānāni bhavanti / tadyathā vṛṛddhipatram, kuśapatram, maṇḍalāyam vrīhimukham, kuṭharākṛti, vatsadantam, utpalapatramē śalākāē sūchīē rampakaścheti / phālajāmavavatāpikāryyākṛtyascheti / ētāndhāgnikarmmavidhāne chatvāri chānyāni śalayaddharaṇāni yathoyogaṁ simhadraṣṭaṁ godhamukhaṁ kaṅkamukhaṁ kuliśamukhaṁ cheti / tisra ēṣiṇayaḥ /

Musical Instruments

Stringed Musical Instruments

The word vīṇā has been used in ancient Indian parlance for string instruments.

Ekatantrī vīṇā is an ancient Indian, one-stringed, fretless, tubular zither with a rounded gourd resonator attached to one end of a long hollowed wooden tube. At the other end, a large wooden peg is inserted, which, in turn, supports a bulky, doubly curved bridge engraved with a thin metallic plate. The playing string, made of animal gut, clamped towards the gourd side of the tube and wrapped around the wooden peg at the other, vibrates while facing a unilateral constraint in the motion due to the finitely curved bridge. A bamboo thread is suitably inserted between the metal plate and the string so the latter makes a grazing contact over the former. The vīṇā is played being held in a slanting position while placing the gourd over the left shoulder and tactfully positioning the bridge on the right heel. The string is plucked with the fingers of the right hand, close to the bridge, whereas the variations in pitch are obtained by sliding a little wooden rod, held in the left hand, along the string. The above description, summarized from the early 13th c. text *Saṅgīta-ratnākara,* is qualitatively consistent with the vīṇā in the hands of the 12th c. Sarasvatī where, even with a broken tube (in the central region), all the essential structural features and the playing mannerisms are unambiguously discernible. The simple structure of the instrument notwithstanding, it is a complicated instrument to play, requiring advanced techniques to produce a rich array of notes all from a single string.

Wind Instruments

Pungi/Been: The Pungi, also called the Been, is one of the oldest wind instruments. It is predominantly played by snake charmers in India and Pakistan. Initially emerging as an accompaniment to Indian folk music, it enjoys an important place in Indian art, culture, and religion, even today.

This instrument includes a mouth-blown air reservoir made from a dried bottle gourd. The neck is curved, and two reed or bamboo pipes are connected at the other end. One of the pipes has seven holes - the player uses this to play the melody. The sound lasts as long as the player does not take pauses.

The pungi, which is still a significant part of Indian folk music and street shows, is also considered to be one of the ways by which one can communicate with the divine.

Flute/Venu/Bansuri: The flute comes   into   the woodwind category. This is a reedless wind instrument, which produces sound from air flow via a small aperture. The person playing a flute is generally called a flautist, flautist, or rarely, fluter or flutenist.

Flutes are ancient musical instruments dating to over 40,000 years ago. These have been an integral part of Indian classical music, both Hindustani and Carnatic. Sri Krishna is always closely associated with the flute.

Indian Flutes: The bamboo flute, primarily used in Indian music, was developed independently of the Western flute. These are simple as compared to the latter and are keyless as well. Indian flutes are mainly of two types, viz., the Venu and the Bansuri.

Venu: The Venu, or Pullanguzhal as it is called, has eight finger holes and is played mainly in South Indian Carnatic music. Most Carnatic musicians use the cross-fingering technique. At the beginning of the 20th Century, these flutes featured only seven finger holes. Sharaba Shastri of the Palladam School developed the standard fingering technique. The quality of the Venu depends much upon the type of bamboo used to create it. Experts agree that the best bamboo can be found in the Nagercoil area in South India.

The Venu finds prominent mention in Indian mythology and folklore. It is listed as one of the three originally indigenous instruments in this country, the others being the veena and the mridangam. The veena-venu-mridangam trio enjoys a significant place in Indian art and culture.

Lord Krishna is usually depicted playing the Venu. This is why he is often addressed as Venugopala. He is believed to be playing the flute to encourage the process of Creation in this world. However, there is no particular name for the flute that the Lord plays.

Bansuri: The word 'bansuri' originates from the 2 Sanskrit words, 'bans' (bamboo) and 'sur' (melody). The typical bansuri is about 14 inches long but could vary between less than 12 inches (muralis) to nearly 40 inches (shankha bansuris). It was traditionally used as a soprano instrument. The bass variety, which Pannalal Ghosh popularized, is now mainly used in Hindustani music. Eventually, what was generally used in folk music came to be used in mainstream Indian classical music.

In order to be suitable for playing, the bamboo used to create a bansuri needs to be thin-walled and straight, with a uniform cross-section. The best quality bamboo can be found in the forests of Assam and Kerala. The bamboo is then seasoned to let it strengthen.

After this, a cork stopper is inserted to block one end, and the blowing hole is burnt. Once all the other holes are burnt in, the bamboo is dipped in a solution of antiseptic oils, cleaned and dried. Its ends are then bound with silk or nylon threads for ornamental and protective purposes.

Longer bansuris with a larger bore feature a lower pitch, and the slimmer, shorter ones are shriller. Since this is a natural woodwind instrument, it is quite delicate and must be maintained carefully.

The bansuri is predominantly used in North Indian or Hindustani music. It is made from a hollow bamboo shaft with six or seven finger holes. Also associated with cowherds and a pastoral environment, this instrument plays a central part in depicting the love between Krishna and Radha and the Rasleela, or the Divine Dance of Radha, Krishna, and the Gopis. Additionally, references to the bansuri can be found in Buddhist paintings from about 100 CE. It is believed that Radha and the Gopis and even animals used to get attracted to the melodies arising from the Lord's flute.

Nadaswaram: The Nadaswaram, referred to as nagaswaram or Nadaswaram, is a double-reed wind instrument that originated in Tamil Nadu and is widely used throughout South India. It is considered the world's loudest acoustic instrument and one of the most challenging instruments to play, similar to the North

Indian Shehnai. It is, however, much longer and more extensive than the latter. Due to its sheer intensity of volume, it is mainly preferred to be an outdoor instrument instead of an indoor, concert-type one.

In Tamil culture, the Nadaswaram is considered a 'mangala vādyam' or auspicious instrument. Hence, it is played in almost all South Indian temples while conducting temple processions, weddings, and other significant ceremonies. The Nadaswaram is mentioned in the ancient treatise, the Silappathikaram. Here, a similar instrument is referred to as the 'Vangiyam'.

This instrument is usually accompanied by a percussive instrument called the 'Thavil'. Another accompaniment with the Nadaswaram is the 'Otthu', which acts as a drone to sustain the sound. The Otthu is played by an assistant junior musician at the leading player's side.

The Nadaswaram comprises three significant parts: kuzhal, thimiru and anasu. Its conical structure gradually curves to enlarge towards the lower end, forming a small speaker-like shape at the end. The top portion has a *mel anaichu* or metal staple, into which the kendai or metallic cylinder, which houses the mouthpiece, is made of reed. Several spare reeds are attached as well. A tiny ivory or horn needle is attached as well in order to clear the reed of saliva or other impurities, thereby allowing for the free flow of sound. The keezh anaichu is a metallic bell forming the instrument's bottom.

Traditionally, the Nadaswaram is made from a tree called aacha. However, nowadays, artisans use various other materials, such as sandalwood, bamboo, copper, brass, ebony, and ivory. Aged wood is considered best to craft a nadaswaram – sometimes, the wood procured from old demolished houses is used.

The Nadaswaram has seven playing holes and five additional ones at the bottom, which is used to manipulate tone. Since seven holes are played with seven fingers, it is also called the 'Ezhil'. It covers a range of two and a half octaves. Semitones and quartertones are produced by adjusting the pressure and intensity of the air flow into the pipe.

Incidentally, there is a smaller version of the Nadaswaram, which is widely used in folk music. This is referred to as the 'mukhavina'.

Shehnai: The Shehnai, also referred to as shenai, shahnai, or simply, maṅgal vādya, is common in North India, Pakistan and Bangladesh. It is a quadruple-reed woodwind instrument with a wooden flared bell at the lower end. Its sound puts it in the category of an auspicious instrument that can sanctify the atmosphere. Hence, like the Nadaswaram, the shehnai is used during temple processions, marriages, and other sacred ceremonies in North India.

The shehnai is a tubular instrument that broadens into a sell at the lower end. It usually has about six to nine holes and uses one set of quadruple reeds. Melodies are played on it by controlling the breath. This instrument admits to a range of two octaves.

Occasionally, two shehnais are tied together and played to create an instrument similar to the Greek Aulos.

Some experts believe that the shehnai evolved from the basic pungi or been. However, others aver that the name 'shehnai' originated from the words 'sur' and 'nal', which mean 'tune' and 'pipe', respectively.

Shehnai players were most popular in North India, Goa, and the Konkan region. The shehnai players, called Vajantri, would serve by playing in the temples of these regions. Each one of them was allotted lands for services thus rendered.

Sāraṅgī: The Sarangi is an ancient Indian bowed, short-necked, fretless instrument. It is also used in Nepal, especially all over the Western part of Nepal; this forms a vital part of Hindustani classical music. Its timbre and resonance make it sound very close to the human voice.

Some experts believe that the word 'sarangi' comes from the words 'saar' (essence) and 'ang' (part of the song or melody). According to folk etymology, the sarangi derived its name from the term, 'sol rang', which means 'a hundred colours'. This indicates its versatility, flexibility, and capability to reproduce just about any melody, gamaka, or mend (ornamentation) the human voice can create. The word 'sarang' has multiple meanings in Sanskrit, so, there is no one definitive reason for the instrument to have gotten this name.

The sarangi is carved from a single block of red cedarwood. It has a box-like shape with three hollow chambers, namely, pet (stomach), chÈti (chest), and magaja (head). It is usually about 2 feet in length and about 6 inches wide. One can find variations in size, however. The vital bridge supports heavy sympathetic steel or brass strings and three main gut strings passing through it. The gut strings are bowed with heavy horsehair bow-the movement of this is controlled with the fingernails, cuticles, and the surrounding flesh.

Extremely difficult to master, this instrument is also painful to play.

Besides these, the sarangi features 35-37 sympathetic strings, divided into 4 'choirs', having two sets of pegs. On the inside is the chromatically tuned row of 15 tarabs, and on the right is the diatonic row of 9 tabs. Each supports a full octave, plus 2-3 extra notes as needed.

Dilruba: The Dilruba, a very ancient musical instrument, is a cross between the sitar and the sarangi. Similar to the Esraj and the Mayuri Veena, the main difference lies in the shape of the resonators and the way the sympathetic strings attach themselves to the instrument. This is a popular instrument in Northwest India and can be found predominantly in Punjab, Uttar Pradesh, and Maharashtra.

The neck of the dilruba consists of about 18 strings. It is tuned similarly to the sitar, and much like the latter, most of the playing is done on one main string. It is a fretted instrument, with the sympathetic strings helping the player sustain the melody's mood as he plays.

The Aruba, being a bowed instrument, is played like the sarangi. It is bowed with the right hand and played using the fingers of the left hand. While some schools mainly use the index finger and the middle finger as the secondary finger, like playing the sitar, other schools prefer using the middle finger, treating the index finger as a subsidiary.

The technique of playing meend is different from the sitar. While the sitar player usually pulls the string along the frets to play semitones and quartertones, the dilruba player merely slides his fingers on the concerned fret to achieve the same sound.

Rāvaṇahatha: The Rāvaṇahatha, alternatively referred to as the Rāvaṇa Hasta Veeṇā, Rāvaṇahattha, Rāvaṇahatta, and Rāvaṇa strong, is one of the most ancient bowed instruments of India. This was popular in Western India and Sri Lanka as well. Incidentally, this instrument is believed to have inspired the Creation of the violin and the viola much later.

The Sinhalese believe that the Rāvaṇahatha originated among their Hela civilization during the rule of their King, Rāvaṇa. In India, this instrument is believed to have existed since 5000 BC. An instrument that closely matches this description can still be found in the remote villages of Rajasthan. It also finds mention in the historical narrations of Rāvaṇa, in the Rāmāyaṇa.

History has it that Rāvaṇa, who was an ardent devotee of Lord Śiva, used to play this instrument in order to please and appease his Lord. History says that after Rāma defeated and killed Rāvaṇa, Hanumān picked up the Rāvaṇahatha and returned with it to North India. From India, it travelled to the Middle East, Europe, and the rest of the Western World, finally giving rise to the violin and the viola in their present form.

A study of the history of Medieval India reveals that the kings and royal princes, who were also patrons of music, helped increase the popularity of the Rāvaṇahasta throughout Rajasthan and Gujarat. Furthermore, the Sangit tradition of Rajasthan encouraged princes, and later, even the royal ladies, to learn to play the instrument.

The bowl of the Rāvaṇahatha, made of coconut shell,

is covered with goat hide. Then, a dandi or a stem made of bamboo is attached to this shell. The two main playing strings are made from steel and horsehair, respectively. These are attached to the instrument with two giant pegs, around which they are wound.

Apart from the two main strings are 8-12 sympathetic strings, which are attached to smaller pegs. The bow is slightly concave and, to its end, are fitted bells meant to provide rhythmic accompaniment..

Percussive Instruments

Mṛdaṅgam: The mṛdaṅgam is a double-sided drum, the body of which is hollowed out from jackfruit wood, which is usually about an inch thick. The two open ends are covered with goatskin and laced with leather straps tied around the circumference of the drum. These straps are held under high tension to stretch the membrane, thus resonating the instrument when struck.

The two membranes are different to produce bass and treble sounds. When struck, the smaller membrane produces a shriller sound, and the wider aperture helps produce lower-pitch sounds. The goatskin covering the smaller end is called 'valanthalai' or 'bala bhaaga' and features a black disk in the centre. This is made of rice flour, ferric oxide powder, and starch. This dark tuning paste is called the 'saadam' or 'karanai' and gives the instrument its metallic sound. The bass end, on the other hand, is known as the 'thoppi' or 'eda bhaaga'. The combination of these two different ends creates the production of the instrument's unique harmonic structure.

The mṛdaṅgam is one of the most ancient percussive

instruments of India and serves as primary rhythmic accompaniment for South Indian Carnatic music, in Bharatanāṭyam and other classical dance forms; and in other forms such as Yakṣagāna as well. In a Carnatic performance, the mridangam is given a significant place for playing solo - referred to as the 'tani' or 'tani avartanam'. Here, the mṛdaṅgist plays his or her solo, alternating with the other upapakkavadyams onstage. After each plays alternating solo pieces that gradually taper in length, they all come together for the grand finale, which ends in a crescendo of sound and rhythm.

Besides India, it is predominantly used in parts of Nepal, Sri Lanka, Malaysia, Singapore, and several other countries of the West. It is often accompanied by upapakkavadyams (secondary instruments) such as the ghatam, kanjira, and morsing.

This instrument was known as the 'tannumai' in early Tamil culture. The word mṛdaṅgam is derived from the two Sanskrit words, 'mṛda' (earth or clay) and aṅga (limb or part). Earlier, mṛdaṅgams were made of the materials mentioned above. Today, it is made mainly from the wood of the jackfruit tree.

The mṛdaṅgam finds a prominent place in several ancient Hindu temples, sculptures, and paintings, especially all over South India. Deities such as Ganesha and Nandi are believed to have played the mridangam and the maddalam to provide rhythmic accompaniment for Śiva's tāṇḍava (Cosmic dance). Hence, this is also termed the Deva Vādyam, the Divine Instrument. Incidentally, this instrument was also played at the start of the war, along with others such as murasu, tudi, and parai, as it was believed to be sacred and had the power to protect the king and his army.

Tabla: Like the m,da×gam, Tabla is a membranophone percussion instrument. It is used as a primary accompaniment in North Indian Hindustani classical music. Besides India, it can be found in Pakistan, Nepal, Afghanistan, Bangladesh, Indonesia, and Sri Lanka. A tabla player is commonly referred to as a tabalchi.

This instrument consists of hand drums of different sizes and, therefore, different timbres. The main drum is a table or dayan (literally, "right") played with the tabalchi's dominant hand. This is conical in shape and is made from wood. Its tightly fitted skin helps produce its distinctive pitch when struck.

The giant drum is lower in pitch and is called bāyān (literally, "left"). It consists of a bowl-shaped metal shell. Because the skin covering it remains loose, the tabalchi can manipulate the sound while playing.

There is mention of the tabla right from the Vedic period in India. Some believe that the mṛdaṅgam inspired its construction and that the latter was halved to create the two drums of the tabla. However, others believe the tabla existed before the mṛdaṅgam came into the scene. The *Saṅgīta Ratnākara,* one of the most important treatises on music, written by Sāraṅgadeva, speaks of an instrument closely resembling the tabla. Ancient Hindu temple carvings in North and South India, dating back to 500 BCE, show double-hand drums looking like the tabla.

The smaller drum is usually made from teak and rosewood, which is hollowed to about half its depth.

This drum is tuned to a pitch that complements the singer's melody. Cylindrical blocks of wood, called ghatta, are inserted between the strap and the shell, thus enabling the player to tune them to his or her preference. Fine-tuning can be achieved by striking vertically on the head using a small hammer-like tool. The giant drum can be made of various materials, including brass, copper, aluminium, or steel. Earlier, wood or clay was used, though this was found to be much less durable.

Both drums are covered with a head made from goat or cow skin. An outer skin is overlaid over the main skin, suppressing some overtones. These two skins are held in place using a complex woven strap braid, which also sustains the tension on the shell. The head of each drum has a tuning paste in the centre, called the syahi, which is made using several layers of paste made from rice or wheat starch and a black powder procured by mixing various ingredients. Each drum is positioned onto a circular ring, called chutta or guddi, for stability and balance. This is made from plant fibre or similar material, wrapped in thick cloth.

Unlike mṛdaṅgists and Pakhawaj players, who play sideways and primarily use the whole palm for playing, Tabalchis use a complex fingertip and hand technique, playing it from the top. This allows the tabalchi to create different types of sounds while playing. The finger technique slightly varies from gharānā (music school) to gharānā.

Dholak: The dholak is a two-headed hand drum used mainly in the folk music of India. It is similar to the larger Punjabi Dhol and the smaller dholki. Besides

India, similar instruments can be found in Pakistan, Suriname, Jamaica, Guyana, the Fiji Islands, Sri Lanka, and the Netherlands. Due to its firm rooting in folk music, it is simple and lacks the advanced tuning and playing techniques of other percussive instruments such as the mridangam, tabla, and pakhawaj. The drum is pitched depending on its size.

The smaller end of the dholak is created with goatskin – this is also the shriller end. The bigger surface, made of buffalo skin, is more bass and gives a lower pitch. The shell is sometimes crafted out of Sheesham wood. However, this instrument is usually made from cheaper wood, such as mango. Dholaks and dholkis in Sri Lanka are made from hollowed-out coconut palm stems.

The dholak is widely used throughout North, Eastern, and Western India during functions and festivities. It is mainly used in the Bhangra, the sprightly folk dance of Punjab; in Lavani, the vibrant folk dance form of Maharashtra; during bhajan, kirtan (devotional song and dance), and qawwali sessions as well.

This instrument prominently features in pre-wedding festivities - children and women dance to ladies singing and playing the dholak. It is also used in filmi sangeet (music) and at baithaks (informal chamber music sessions).

Pañchavādyam: Pañchavādyam, which means "orchestra of five instruments", is a temple art form that developed in Kerala. The five instruments included in this unique ensemble are timila, maddalam, ilathalam, idakka, and kombu. Of these, only the kombu is a wind instrument. All the rest are percussive instruments.

The pañchavādyam features a pyramid-like rhythmic structure that steadily increases in tempo, with a proportionate decrease in the number of beats per tala cycle. Though it is a temple art form, this does not relate to any particular temple ritual - the artists, too, use personal improvisation to add more shades to the performance.

Appendix

समरांगणसूत्रधारे यन्त्रविधानं नामैकत्रिंशोऽध्यायः ।

samarāṁgaṇasūtradhāre yantravidhānaṁ nāmaikatrimśo'dhyāyaḥ ।

भ्राम्यद्दिनेशशशिमण्डलचक्रशत्सं(स्त)
मेतज्जगच्चितयययन्त्रमलक्ष्यमध्यम् ।

भूतानि बीजमखिलान्यपि सम्प्रकल्प्य
यः सन्ततं भ्रमयति स्मरजित् स वोऽव्यात ॥1॥

bhrāmyaddineśaśaśimaṇḍalachakraśatsaṁṁ(sta)
metajjagattritayayantramalakṣayamadhyam ।

bhūtāni bījamakhilānyapi samprakalpya
yaḥ santataṁ bhramayati smarajit sa vo'vyāta ॥1॥

यन्त्राध्यायमथ ब्रूमो यथावत् प्रकमागतम् ।
धर्मार्थकाममोक्षाणां यदेकमिह कारणम् ॥2॥

yantrādhyāyamatha brūmo yathāvat prakamāgatam ।
dharmārthakāmamokṣāṇāṁ yadekamiha kāraṇam ॥2॥

यद्च्छया प्रवृत्तनि भूतानि स्वेन वर्त्मना ।
नियम्यास्मिन् नयति यत् तद् यन्त्रमिति कीर्तितम् ॥3॥

yaddachchhayā pravṛttani bhūtāni svena vartmanā ।
niyamyāsmin nayati yat tad yantramiti kīrtitam ॥3॥

स्वरसेन प्रवृत्तानि भूतानि स्वमनीषया
कृतं यस्माद् यमयति तद्वा यन्त्रमिति स्मृतम् ॥4॥

svarasena pravṛttāni bhūtāni svamanīṣayā
kṛtaṁ yasmād yamayati tadvā yantramiti smṛtam ॥4॥

तस्य बीजं चतुर्धा स्यात् क्षितिरापोऽनाऽनिलः ।

आश्रयत्वेन चैतेषां विजयदप्युपयुज्यते ॥5 ॥

tasya bījaṁ chaturdhā syāt kṣitirāpo'nā'nilaḥ /
āśrayatvena chaiteṣāṁ vijayadapyupayujyate ॥5 ॥

भिन्नः सूतश्चकै(यै) रुक्तस्ते च सम्यङ् न जानते ।
प्रकृत्या पार्थिवः सूतल(सत्र) यात् तत्र क्रिया भवेत् ॥6 ॥

bhinnaḥ sūtaśchakai(yai) ruktaste cha samyaṅ na jānate /
prakṛtyā pārthivaḥ sūtala(satra) yāt tatra kriyā bhavet ॥6 ॥

पार्थिवत्वादयमतो न कदाचिद् विभिद्यते ।
द्रव्यत्वादग्निजत्व हि यद्यस्य परिकल्पयते ॥7 ॥

pārthivatvādayamato na kadāchid vibhidyate /
dravyatvādagnijatva hi yadyasya parikalpayate ॥7 ॥

तदा विरोधो नैवास्य पावकेनोपपद्यते ।
गन्धाद वह्नेविरोधांच स्थिता पार्थिववता बलात् ॥8 ॥

tadā virodho naivāsya pāvakenopapadyate /
gandhāda vahnevirodhāṁcha sthitā pārthivavatā balāt ॥8 ॥

आत्मैव बीजं सर्वेषां प्रत्येकमराण्यपि ।
एवं भेदा भवन्त्येषां भूयांसः सङ्करान्मिथः ॥9 ॥

ātmaiva bījaṁ sarvaṣāṁ pratyekamarāṇyapi /
ēvaṁ bhedā bhavantyeṣāṁ bhūyāṁsaḥ saṅkarānmithaḥ ॥9 ॥

स्वयंवाह्यमिहोत्कृष्टं हीनं स्यादितरत् त्रयम् ।
अन्यदन्तरितं वाह्यं वाह्यमन्यत् त्वदूरतः ॥ 10 ॥

svayaṁvāhyamihotkṛṣṭaṁ hīnaṁ syāditarat trayam /
anyadantaritaṁ vāhyaṁ vāhyamanyat tvadūrataḥ ॥ 10 ॥

स्वयं वाह्यमिहोत्कृष्टं हीनं स्यादितरत् त्रयम् ॥
तेषु शंसन्ति दूरस्थलक्ष्यं निकटस्थितम् ॥11 ॥

svayaṁ vāhyamihotkṛṣṭaṁ hīnaṁ syāditarat trayam ॥
teṣu śaṁsanti dūrasthalakṣayaṁ nikaṭasthitam ॥11 ॥

यद्यु(दु) त्पन्नमलक्ष्यं यदेकं बहुषु साधकम् ।
तदन्यदपि शंसन्ति यस्माद् विस्मयकृन्नृणाम् ॥12 ॥

yadyu(du) tpannamalakṣayaṁ yadekaṁ bahuṣu sādhakam |
tadanyadapi śaṁsanti yasmād vismayakṛnnṛṇām ||12 ||

एका स्वीया गतिश्च वाक्तह्येऽन्या वाहकाश्रिता ।
अरघट्टश्रिते कीटे दृश्यते द्वयमप्यदः ॥13 ॥

ēkā svīyā gatiścha vāktahye'nyā vāhakāśritā |
araghaṭṭaśrite kīṭe dṛśyate dvayamapyadaḥ ||13 ||

इत्थं गतिद्वयवशाद् वैचित्र्यं कल्पयेत् स्वयम् ।
अलक्षता विचित्रत्वं यस्माद् यन्त्रेषु शस्यते ॥14 ॥

ittham gatidvayavaśād vaichitrayaṁ kalpayet svayam |
alakṣatā vichitratvaṁ yasmād yantreṣu śasyate ||14 ||

अन्यत् स्यादन्तरात्रे(प्रे)र्यं द्वितीयं मध्यमं त्विदम् ।
द्वयत्रयदियोगेन चतुर्णामपि योगतः ॥ |15 ॥

anyat syādantarātpre(pre)ryaṁ dvitīyaṁ madhyamaṁ
tvidam |
dvayatrayadiyogena chaturṇāmapi yogataḥ || |15 ||

आंशांशिभावाद् भूतानां सङ्ख्यैषामतिरिच्यते ।
यः सम्यगेतज्ज्ञानाति स पुमान् भवति प्रियः ॥16 ॥

āṁsāṁśibhāvād bhūtānāṁ saṅkhyaiṣāmatirichyate |
yaḥ samyagetajjānāti sa pumān bhavati priyaḥ ||16 ||

प्रमदानां नृपाणां च प्रज्ञानां च मतस्य च ।
लाभं ख्यातिं च पूजां च यशो मानं धनानि च ॥17 ॥

pramadānāṁ nṛpāṇāṁ cha prajñānāṁ cha matasya cha |
lābhaṁ khyātiṁ cha pūjāṁ cha yaśo mānaṁ dhanāni
cha ||17 ||

प्राप्नोति किं किं न पुमान् य इदं वेत्त तत्त्वतः ।
गृहमेकं विलासनामाश्चर्यस्य परं पदम् ॥18 ॥

prāpnoti kiṁ kiṁ na pumān ya idaṁ vetta tattvataḥ |
gṛhamekaṁ vilāsanāmāścharyasya paraṁ padam ||18 ||

रतेरावासभवनं विस्मयस्यैकमास्पदम् ।

यथावद् देवतादीनां रूपचेष्टादिदर्शनात् ॥19॥

raterāvāsabhavanaṁ vismayasyaikamāspadam |
yathāvad devatādīnāṁ rūpacheṣṭādidarśanāt ॥19॥

तास्तुष्यन्त्यथ तत्तुष्टिः पूर्वैधर्मः प्रकीर्तितः ।
नृपादितोषादर्थः स्यादर्थे कामः प्रतिष्ठित ॥20॥

tāstuṣyantyatha tattuṣṭiḥ pūrvaidharmaḥ prakīrtitaḥ |
nṛpāditoṣādarthaḥ syādarthe kāmaḥ pratiṣṭhita ॥20॥

वित्तैक्यादस्य निष्पत्तिर्मोक्षश्चास्मान्न दुर्लभः ।
पार्थिवं पार्थवैर्बीजैः पार्थवं जलजन्मभिः ॥21॥

vittaikyādasya niṣpattirmokṣaśchāsmānna durlabhaḥ |
pārthivaṁ pārthavairbījaiḥ pārthavaṁ jalajanmabhiḥ ॥21॥

तदेव तेजोजनितैस्तदेव मरुदुद्भवैः ।
आप्यमाप्यैस्तथा बीजैरानलरानिलैरपि ॥22॥

tadeva tejojanitaistadeva marududbhavaiḥ |
āpyamāpyaistathā bījairānalarānilairapi ॥22॥

वह्निजैश्च मरुज्जातैः पार्थिवैर्वारुणैरपि ।
मारुतं मारुतैराप्यैः पार्थिवैरानलैस्तथा ॥ 23 ॥

vahnijaiścha marujjātaiḥ pārthivairvāruṇairapi |
mārutaṁ mārutairāpyaiḥ pārthivairānalaistathā ॥ 23 ॥

वह्निजातेऽपि बीजं स्यात् सूतः सोऽपि च वान(नि) ।
पार्थिवानां भवेद् बीजमाप्यानामपि वारणे(रुणम्) ॥ 24 ॥

vahnijāte'pi bījaṁ syāt sūtaḥ so'pi cha vāna(ni) |
pārthivānāṁ bhaved bījamāpyānāmapi vāraṇe(ruṇam) ॥ 24

इति बीजानि सर्वेषां कीर्तितान्यखिलान्यपि ।
कुडयंकरणसूत्राणि भारगोलकपीडनम् ॥25॥

iti bījāni sarveṣāṁ kīrtitānyakhilānyapi |
kuḍayaṁkaraṇasūtrāṇi bhāragolakapīḍanam ॥25॥

लम्बनं लम्बकारे च चक्राणि विविधान्यपि ।
अयस्ताम्रं च तारं च त्रपु संवित्रमर्दने ॥26॥

lambanaṁ lambakāre cha chakrāṇi vividhānyapi |
ayastāmraṁ cha tāraṁ cha trapu saṁvitpramardane ||26 ||

काष्ठं च चर्म वस्त्रं च स्वबीजेषु प्रयुज्यते ।
उर्दकः कर्तरो यष्टिश्चक्रं भ्रमरकस्था ॥27 ॥

kāṣṭhaṁ cha charma vastraṁ cha svabījeṣu prayujyate |
urdakaḥ kartaro yaṣṭiśchakraṁ bhramarakasthā ||27 ||

श्रृङ्गावली च नाराचः स्वबीजान्यौर्वरे विदुः ।
ताप उत्तेजनं स्तोभः क्षेभश्च जलसङ्गजः ॥28 ॥

śṛṅgāvalī cha nārāchaḥ svabījānyaurvare viduḥ |
tāpa uttejanaṁ stobhaḥ kṣebhaścha jalasaṅgajaḥ ||28 ||

एवमाद्यग्निबीजानि पार्थिवस्य प्रचक्षते ।
धारा च जलभारश्च पयसो भ्रमणं तथा ॥29 ॥

ēvamādyagnibījāni pārthivasya prachakṣate |
dhārā cha jalabhāraścha payaso bhramaṇaṁ tathā ||29 ||

एवमदीनि भूजस्य जलजानि प्रचक्षते ।
यथोच्छायो यथाधिक्यं यथा नीरन्ध्रतापि च ॥30 ॥

ēvamadīni bhūjasya jalajāni prachakṣate |
yathochchhāyo yathādhikyaṁ yathā nīrandhratāpi
cha ||30 ||

अत्यन्तमूर्ध्वगामित्वं स्वबीजान्ययसस्तथा ।
मरुत् स्वभावजो गाढै ग्रहकैश्च प्रतीप्सितः ॥31 ॥

atyantamūrdhvagāmitvaṁ svabījānyayasastathā |
marut svabhāvajo gāḍhai rgrāhakaiścha pratīpsitaḥ ||31 ||

दृत्याद्यैर्वीजनाद्यैश्च गजकर्णादिभिः कृतः ।
(छ) चाणितो गालितश्चायं बीजं भवति भूभवे ॥ 32 ॥

dṛtyādyairvījanādyaiścha gajakarṇādibhiḥ kṛtaḥ |
(chha) chāṇito gālitaśchāyaṁ bījaṁ bhavati bhūbhave || 32

काष्ठंभृ(कृ) त्तिश्च लोहं च जलजे पार्थिवं भवेत् ।
अन्यदम्भस्तदप्यस्तु तिर्यगूर्ध्वमधस्तथा ॥33 ॥

kāṣṭambhṛ(kṛ) ttiścha loham cha jalaje pārthivam bhavet /
anyadambhastadapyastu tiryagūrdhvamadhastathā //33 //

बीजं स्वकीयं भवति यन्त्रेषु जलजन्मसु।
तापाद्यं पूर्वकथितं वह्निजं जलजे भवेत्॥34॥

bījam svakīyam bhavati yantreṣu jalajanmasu /
tāpādyam pūrvakathitam vahnijam jalaje bhavet //34 //

सङ्गृहीतश्च दत्तश्च पूरितः प्रतिनोदितः।
मरुद् बीजत्वमायाति यन्त्रेषु जलजन्मसु॥35॥

saṅgṛhītaścha dattaścha pūritaḥ pratinoditaḥ /
marud bījatvamāyāti yantreṣu jalajanmasu //35 //

वह्निजातेषु मृत्ताम्रलोहरुक्मादि तद्ग्रहे।
पार्थिवं कथयन्तीह बीजं बीजविचक्षणाः॥36॥

vahnijāteṣu mṛttāmraloharukmādi tadgrahe /
pārthivam kathayantīha bījam bījavichakṣaṇāḥ //36 //

वह्नेर्वह्निभिर्विवेद् बीजमाप आपस्तथा भवेत्।
आद्यैर्दृत्यादिभिः प्रोक्तैर्मरुद् गच्छति बीजताम्॥37॥

vahnervahnarbhived bījamāpa āpastathā bhavet /
ādyairdṛtyādibhiḥ proktairmarud gachchhati bījatām //37 //

प्रत्येकं च जनकं प्रेरकं ग्राहकं तथा।
सङ्ग्राहकं च भूजातं बीजं स्यादनिलोद्भवैः॥38॥

pratyekam cha janakam prerakam grāhakam tathā /
saṅgrāhakam cha bhūjātam bījam syādanilodbhavaiḥ //38 //

प्रेरणं चाभिघातश्च विवर्तो भ्रमणं तथा।
जलजं मारुतोत्थेषु बीजं स्यादिति सम्तम्॥39॥

preraṇam chābhighātaścha vivarto bhramaṇam tathā /
jalajam mārutottheṣu bījam syāditi samtam //39 //

सङ्गृहीतस्य तापाद्यैर्यानि पावकजन्मनि।
प्रकीर्तितानि तान्येव भवन्ति पवनोद्भवैः॥40॥

saṅgṛhītasya tāpādyairyāni pāvakajanmani /

prakīrtitāni tānyeva bhavanti pavanodbhavaiḥ ||40 ||

प्रेरितः सङ्गृहीतश्च जनितश्च समीरणः ।
आत्मनो बीजतां गच्छत्येवमन्यत् प्रकल्पयेत् ॥41 ॥

preritaḥ saṅgṛhītaścha janitāścha samīraṇaḥ |
ātmano bījatāṁ gachchhatyevamanyat prakalpayet ||41 ||

भूतमेकमिहोद्रिक्तमन्यद्धीनं तोऽधिकम् ।
अन्यद्धीनतरं चान्यदेवंप्रायैर्विकल्पितैः ॥42 ॥

bhūtamekamihodriktamanyaddhīnaṁ to'dhikam |
anyaddhīnataraṁ chānyadevaṁprāyairvikalpitaiḥ ||42 ||

नाना भेदा भवन्त्येषां कस्तान् कात्स्र्येन वक्ष्यति ।
निष्क्रिया भूः क्रिया त्वंशे शेषेषु सहजा त्रिषु ॥43 ॥

nānā bhedā bhavantyeṣāṁ kastān kārtsnyena vakṣayati |
niṣkriyā bhūḥ kriyā tvaṁśe śeṣeṣu sahajā triṣu ||43 ||

अतः प्रायेण सा जन्या क्षितावेव प्रयत्नतः ।
साध्यस्य रूपशवतः सन्निवेशो यतो भवेत् ॥44 ॥

ataḥ prāyeṇa sā janyā kṣitāveva prayatnataḥ |
sādhyasya rūpaśavataḥ sanniveśo yato bhavet ||44 ||

यन्त्राणामाकृतिस्तेन निर्णेंतु नैव शक्यते ।
यथावद्बीजसंयोगः सौश्लिष्टयं श्लक्ष्णतापि च ॥45 ॥

yantrāṇāmākṛtistena nirṇeṁtu naiva śakyate |
yathāvadbījasaṁyogaḥ sauśliṣṭayaṁ ślakṣaṇatāpi cha ||45 ||

अलक्षता निर्वहणं लघुत्वं शब्दहीनता ।
शब्दे साध्ये तदाधिक्यमशैथिल्यमगाढता ॥46 ॥

alakṣatā nirvahaṇaṁ laghutvaṁ śabdahīnatā |
śabde sādhye tadādhikyamaśaithilyamagāḍhatā ||46 ||

वहनीषु समस्तासु सौश्लिष्टयं चास्खलद्गति ।
यथाभीष्टार्थकारित्वं लयतालानुगामिता ॥47 ॥

vahanīṣu samastāsu sauśliṣṭayaṁ chāskhladgati |
yathābhīṣṭārthakāritvaṁ layatālānugāmitā ||47 ||

इष्टकालेऽर्थदर्शिलं पुनः सम्यक्त्त्वसंवृतिः ॥
अनुलबणलं ताद्रूप्यं दाढर्ये मसृणता तथा ॥48॥

iṣṭakāle'rthadarśitvaṁ punaḥ samyaktvasaṁvṛtiḥ ॥
anulabaṇatvaṁ tādrūpyaṁ dāḍharye masṛṇatā tathā ॥48॥

चिरकालसहलं च यन्त्रस्यैते गुणाः स्मृताः ।
एकं बहूनि चलयेद् बहुभिश्चाल्यतेऽपरम् ॥49॥

chirakālasahatvaṁ cha yantrasyaite guṇāḥ smṛtāḥ ।
ēkaṁ bahūni chalayed bahubhiśchālyate'param ॥49॥

सुश्लिष्टत्वमलक्षलं यन्त्राणां परमो गुणः ।
अथ कर्माणि यन्त्राणां विचित्राणि यथाविधि ॥50॥

suśliṣṭatvamalakṣatvaṁ yantrāṇāṁ paramo guṇaḥ ।
atha karmāṇi yantrāṇām vichitrāṇi yathāvidhi ॥50॥

न विस्तरान्न सङ्क्षेपात् साम्प्रतं संप्रचक्ष्महे ।
कस्यचित् सा क्रिया साध्या कालः कस्यापि कस्यचित् ॥51॥

na vistarānna saṅkṣepāt sāmpratam samprachakṣamahe ।
kasyachit sā kriyā sādhyā kālaḥ kasyāpi kasyachit ॥51॥

शब्दः कस्यापि चोच्छ्रायो रूपस्पर्शौ च कस्यचित् ।
क्रियास्तु कार्यस्य वशादनन्ताः परिकीर्तिताः ॥52॥

śabdaḥ kasyāpi chochchhrāyo rūpasparśau cha kasyachit ।
kriyāstu kāryasya vaśādanantāḥ parikīrtitāḥ ॥52॥

तीर्यंगूध्वमधः पृष्ठे पुरतः पार्श्वयोरपि ।
गमनं सरणं पात इति भेदाः क्रियोद्भवाः ॥53॥

tīryaṁgūdhvamadhaḥ pṛṣṭhe purataḥ pārśvayorapi ।
gamanaṁ saraṇaṁ pāta iti bhedāḥ kriyodbhavāḥ ॥53॥

कालो मुहूर्तकाष्ठाद्यैर्भिन्नो भेदैरनेकधा ।
शब्दो विचित्रः सुखदो रतिकृद् भीषणस्तथा ॥54॥

kālo muhūrtakāṣṭhādyairbhinno bhedairanekadhā ।
śabdo vichitraḥ sukhado ratikṛd bhīṣaṇastathā ॥54॥

उच्छ्रायस्तु जलस्य स्यात् क्रचिद् भूजेऽपि शस्यते ।

गीतं नृत्यं च वाद्यं च पटहो वंश एव च ॥55॥

uchchhrāyastu jalasya syāt kvachid bhūje'pi śasyate |
gītam nṛtyam cha vādyam cha paṭaho vamśa ēva cha ॥55॥

वीणा च कांस्यतालश्च तृमिला करटापि च ।
यत्किंचदन्यदप्यत्र वादित्रादि विभाव्यते ॥56॥

vīṇā cha kāṃsyatālaścha tṛmilā karaṭāpi cha |
yatkiimchadanyadapyatra vāditrādi vibhāvyate ॥56॥

समस्तमपि तद् यन्त्राज्जायते कल्पनावशात् ।
नृत्ये तु नाटकं चोक्षस्ताण्डवं लास्यमेव च ॥57॥

samastamapi tad yantrājjāyate kalpanāvaśāt |
nṛtye tu nāṭakam chokṣastāṇḍavam lāsyameva cha ॥57॥

राजमार्गश्च देशी च यन्त्रात् सर्वं प्रसिध्यति ।
तथा जात्यनुगाश्चेष्टा विरुद्धा यास्तु जातितः ॥58॥

rājamārgaścha deśī cha yantrāt sarvam prasidhyati |
tathā jātyanugāścheṣṭā viruddhā yāstu jātitaḥ ॥58॥

ताः सर्वा पि सिध्यन्ति सम्यग्यन्त्रस्य साधनात् ।
भूचराणां गतिर्व्योम्नि भूमौ व्योमचरागमः ॥59॥

tāḥ sarvā pi sidhyanti samyagyantrasya sādhanāt |
bhūcharāṇām gatirvyomni bhūmau
vyomacharāgamaḥ ॥59॥

चेष्टितान्यपि मर्त्यानां तथा भूमिस्पृशामिव ।
जायन्ते यन्त्रनिर्माणाद् विविधानीप्सितानि च ॥60॥

cheṣṭitānyapi martyānām tathā bhūmispṛśāmiva |
jāyante yantranirmāṇād vividhānīpsitāni cha ॥60॥

यथासुरा जिता दैवैर्यथा निर्मथितोऽम्बुधिः ।
हिरण्यकशिपुर्दैत्यो नृसिंहेन हतो यथा ॥61॥

yathāāsurā jitā daivairyathā nirmathito'mbudhiḥ |
hiraṇyakaśipurdaityo nṛsimhena hato yathā ॥61॥

धावनं हस्तियुद्धं च गजानामगडोऽपि च ।

नानाप्रकार(रा) या चेष्टा नानाधारागृहाणि च ॥62॥

dhāvanaṁ hastiyuddhaṁ cha gajānāmagaḍo'pi cha |
nānāprakāra(rā) yā cheṣṭā nānādhārāgrhāṇi cha ॥62॥

दोलाकेल्यो विचित्रश्च तथा रतिगृहाणि च।
चित्रा सेन(ना) च कुटयश्च स्वयंवाहकसेवकाः ॥63॥

dolākelyo vichitraścha tathā ratigrhāṇi cha |
chitrā sena(nā) cha kuṭayaścha svayaṁvāhakasevakāḥ ॥63॥

सभाश्च विविधाकाराः सत्या मायाः प्रकल्पिताः।
एवंप्रायाणि चान्यानि यन्त्रात् सिध्यन्ति कल्पनात् ॥64॥

sabhāścha vividhākārāḥ satyā māyāḥ prakalpitāḥ |
ēvaṁprāyāṇi chānyāni yantrāt sidhyanti kalpanāt ॥64॥

विधाय भूमिकाः पंच शय्या ल्वा दिभुवि स्थिता।
प्रतिप्रहरमन्यासु सर्पन्ती याति पंचमीम् ॥65॥

vidhāya bhūmikāḥ paṁcha śayyā tvā dibhuvi sthitā |
pratipraharamanyāsu sarpantī yāti paṁchamīm ॥65॥

एवंप्रायाणि चित्राणि सम्यक् सिध्यन्ति यन्त्रतः।
क्रमेण त्रिशतावर्तं स्थाले दन्ता भ्रमन्त्यसौ ॥66॥

ēvaṁprāyāṇi chitrāṇi samyak sidhyanti yantrataḥ |
krameṇa triśatāvartaṁ sthāle dantā bhramantyasau ॥66॥

तन्मध्ये पुत्रिका क्लृप्ता प्रति नाडिं प्रबांधयेत्।
वह्नेश्च दर्शनं तोये वह्निमध्याज्जलोद्ग्तिः ॥67॥

tanmadhye putrikā klṛptā prati nāḍiṁ prabāṁdhayet |
vahneścha darśanaṁ toye vahnimadhyājjalodgatiḥ ॥67॥

अवस्तुतोऽपि वस्तुलं वस्तुतोऽपि तथान्यथा।
निःश्वासेन वियद् याति श्वासेनायाति मेदिनीम् ॥68॥

avastuto'pi vastutvaṁ vastuto'pi tathānyathā |
niḥśvāsena viyad yāti śvāsenāyāti medinīm ॥68॥

क्षीरोदमध्यगा शय्या प्रतीष्ठाधः फणाभृता।
गोलश्च सूति(च)विहितः सूर्यादीनां प्रदक्षिणम् ॥69॥

kṣīrodamadhyagā śayyā pratiṣṭādhaḥ phaṇābhṛtā /
golaścha sūti(cha)vihitaḥ sūryādīnāṁ pradakṣiāṇam //69//

परिभ्राम्यत्यहोरात्रं ग्रहाणां दर्शयन् गतिम्।
गजादिरूपे रथिकरूपतां गमितः पुमान्॥70॥

paribhrāmyatyahorātraṁ grahāṇāṁ darśayan gatim /
gajādirūpe rathikarūpatāṁ gamitaḥ pumān //70//

भ्रान्त्वा नाडिकया तस्याः पर्यन्ते हन्ति भो(यो) जनम्।
दीपिकापुत्रिका क्लृप्ता क्षीणं क्षीणं प्रयच्छति॥71॥

bhrāntvā nāḍikayā tasyāḥ paryante hanti bho(yo) janam /
dīpikāputrikā klṛptā kṣīṇaṁ kṣīṇaṁ prayachchhati //71//

दीपे तैलं प्रनृत्यन्ती तालगन्त्या प्रदक्षिणम्।
यावत् प्रदीयते वारि तावत् पिबति सन्ततम्॥72॥

dīpe tailaṁ pranṛtyantī tālagantyā pradakṣiṇam /
yāvat pradīyate vāri tāvat pibati santatam //72//

यन्त्रेण कल्पितो हस्ती न तद् गच्छत् प्रतीयते।
शुकाद्याः पक्षिणः क्लृप्तास्तालस्यानुगमान्मुहुः॥73॥

yantreṇa kalpito hastī na tad gachchhat pratīyate /
śukādyāḥ pakṣiṇaḥ klṛptāstālasyānugamānmuhuḥ //73//

जनस्य विस्यकृतो नृत्यन्ति च पठन्ति च।
पुत्रिका वा गजेन्द्रो वा तुरगो कर्मटोऽपि वा॥74॥

janasya visyakṛto nṛtyanti cha paṭhanti cha /
putrikā vā gajendro vā turago karmaṭo'pi vā //74//

वलनैर्वर्तनैनैर्नृत्यंस्तालेन हरते मनः॥
येनैव वत्म्रना क्षेत्रं ध्रियते तेन तत्पयः॥75॥

valanairvartanairnṛtyaṁstālena harate manaḥ //
yenaiva vatmranā kṣetraṁ dhriyate tena tatpayaḥ //75//

यात्यायाति पुनस्तद्वद् गर्तात् पुष्करिणीष्वपि।
फलके कानि(?)तिष्ठन्ति धावन्त्यनुमतानि च॥76॥

yātyāyāti punastadvad gartāt puṣkariṇīṣvapi /

phalake kāni(?)tiṣṭhanti dhāvantyanumatāni cha ॥76॥

घातां(तं) ददति युध्यन्ते निर्यान्त्यश्रमनावृतम् ।
नृत्यन्ति गायन्ति तथा वंशादीन् वादयन्ति च ॥77॥

*ghātāṁ(taṁ) dadati yudhyante niryāntyaśramanāvṛtam ।
nṛtyanti gāyanti tathā vaṁśādīn vādayanti cha ॥77॥*

निरुद्धमुक्तस्य वशान्मरुतो यन्त्रभङ्गिभिः ।
याश्रेष्ठा दिव्यमानुष्यस्ता एवान्न न केवलम् ॥78॥

*niruddhamuktasya vaśānmaruto yantrabhaṅgibhiḥ ।
yāśchesṭā divyamānuṣyastā ēvānna na kevalam ॥78॥*

दुष्करं यद्यदन्यच्च तत्तद् यन्त्रात् प्रसिध्यति ।
यन्त्राणां घटना नोक्ता गुप्तर्थं नाज्ञतावशात् ॥79॥

*duṣkaraṁ yadyadanyachcha tattad yantrāt prasidhyati ।
yantrāṇāṁ ghaṭanā noktā guptarthaṁ nājñatāvaśāt ॥79॥*

तन्न हेतुरयं ज्ञेयो व्यक्ता नैते फलप्रदाः ।
कथितान्यत्र बीजानि यन्त्राणां घटना न यत् ॥80॥

*tanna heturayaṁ jñeyo vyaktā naite phalapradāḥ ।
kathitānyatra bījāni yantrāṇāṁ ghaṭanā na yat ॥80॥*

तस्माद् व्यक्तीकृतेष्वेषु न स्यात् स्वार्थो न कौतुकम् ।
वस्तुतः कथितं सर्वं बीजानामिह कीर्तनात् ॥81॥

*tasmād vyaktīkṛteṣveṣu na syāt svārtho na kautukam ।
vastutaḥ kathitaṁ sarvam bījānāmiha kīrtanāt ॥81॥*

अभ्यूह्यं स्वधिया प्राज्ञैर्यन्त्राणां कर्म यद् यथा ।
यन्त्राणि यानि दृष्टानि कीर्तितान्यत्र तान्यपि ॥82॥

*abhyūhyaṁ svadhiyā prājñairyantrāṇāṁ karma yad yathā ।
yantrāṇi yāni dṛṣṭāni kīrtitānyatra tānyapi ॥82॥*

नन्द्यानि यस्मात् तान्यातो विज्ञेयान्युपदेशतः ॥
एतत् स्वबुद्ध्यैवास्माभिः समग्रमपि कल्पितम् ॥83॥

*nandyāni yasmāt tānyāto vijñeyānyupadeśataḥ ॥
ētat svabuddhyaivāsmābhiḥ samagramapi kalpitam ॥83॥*

अग्रतश्च पुनर्ब्रूमः कथितं यत् पुरातनैः ।
बीजं चतुर्विधमिह प्रवदन्ति यन्त्रे-
ष्वम्भोग्निभूमिपवनैर्निहितैर्यथावत् ।
प्रत्येकतो बहुविधं हि विभागतः स्या-
न्मिश्रैर्गुणैः पुनरिदं गणनामपास्येत् ॥84॥

agrataścha punarbrūmaḥ kathitaṁ yat purātanaiḥ |
bījaṁ chaturvidhamiha pravadanti yantre-
ṣvambhognibhūmipavanairnihitairyathāvat |
pratyekato bahuvidhaṁ hi vibhāgataḥ syā-
nmiśrairguṇaiḥ punaridaṁ gaṇanāmapāsyet ॥84॥

किमेतस्मादन्यद भवति भुवने वित्रमपरं
किमन्यद् वा तुच्छ्यै भवति किमु वा कौतुककरम् ।
किमन्यद् वा कीर्त्तेर्भवनमपरं कामसदनं
किस्मात् पुण्यं वा किमिव च परीतापशमनम् ॥85॥

kimetasmādanyada bhavati bhuvane vitramaparaṁ
kimanyad vā tuṣṭhyai bhavati kimu vā kautukakaram |
kimanyad vā kīrtterbhavanamaparaṁ kāmasadanaṁ
kismāt puṇyaṁ vā kimiva cha parītāpaśamanam ॥85॥

एतेऽत्यर्थं प्रीतिदा बीजयोगाः
संजायन्ते योजिताः सूत्रधारैः ।
भ्रान्त्या नान्यश्चित्रकृद् दारुक्लृप्तं
चक्रं दोलाद्यं पुनः पंचमं तत् ॥86॥

ēte'tyarthaṁ prītidā bījayogāḥ
saṁjāyante yojitāḥ sūtradhāraiḥ |
bhrāntyā nānyaśchitrakṛd dāruklṛptaṁ
chakraṁ dolādyaṁ punaḥ paṁchamaṁ tat ॥86॥

पारम्पर्यं कौशलं सौपदेशं
शास्त्राभ्यासो वास्तुकर्मोद्यमो धीः ।
सामग्रीयं निर्मला यस्य सोऽस्मिं-
श्चित्राण्येवं वेत्ति यन्त्राणि कर्तुम् ॥87॥

pāramparyaṁ kauśalaṁ saupadeśaṁ

śāstrābhyāso vāstukarmodyamo dhīḥ |
sāmagrīyaṁ nirmalā yasya so'smiṁ-
śchitrāṇyevaṁ vetti yantrāṇi kartum ||87 ||

चित्रैर्युक्तं ये गुणः पंचरूपं
ज्ञानन्त्येनं यन्त्रशास्त्रधिकारम् ।
ये वा कृत्स्नं योजयन्तेऽत्र
सम्यक् तेषां कीर्त्तिर्द्यां भुवं चावृणोति ॥88 ॥

chitrairyuktaṁ ye guṇaḥ paṁcharūpaṁ
jñānantyenaṁ yantraśāstradhikāram |
ye vā kṛtsnaṁ yojayante'tra
samyak teṣāṁ kīrttirdyāṁ bhuvaṁ chāvṛṇoti ||88 ||

अङ्गुलेन मितमङ्गुलपादेनोच्छ्रितं द्विपुटकं तनुवृत्तम् ।
संविधेयमृजु मध्यगरन्ध्रं श्लिष्टसन्धि दृढताम्रमयं तत् ॥ ।89 ॥

anagulena mitamangulapādenochchhritaṁ dviputakaṁ
tanuvṛttam |
saṁvidheyamṛju madhyagarandhraṁ śliṣṭasandhi
dṛḍhatāmramayaṁ tat || |89 ||

दारवेषु विहगेषु तदन्तः क्षिप्तमुद्गतसमीरवशेन ।
आतनोति विचलन्मृदुशब्दं शृण्वतां भवति चित्रकरं च ॥90 ॥

dāraveṣu vihageṣu tadantaḥ kṣiptamudgatasamīravaśena |
ātanoti vichalanmṛduśabdaṁ śṛṇvatāṁ bhavati chitrakaraṁ
cha ||90 ||

सुश्लिष्टखण्डद्वितयेन कृत्वा सरन्ध्रमन्तर्मुजानुकारम् ।
ग्रस्तं तथा कुण्डलयोर्युगेन मध्ये पुटं तस्य मृदु प्रदेयम् ॥91 ॥

suśliṣṭakhaṇḍadvitayena kṛtvā
sarandhramantarmujānukāram |
grastaṁ tathā kuṇḍalayoryugena madhye puṭaṁ tasya
mṛdu pradeyam ||91 ||

पूर्वोक्तयन्त्रे विधिनोदरेऽस्य क्षिप्तेऽथ शय्यातलसंस्थमेतत् ।
ध्वनिं ततः संचलनादनङ्गक्रीडारसोल्लासकरं करोति ॥92 ॥

pūrvoktayantre vidhinodare'sya kṣipte'tha
śayyātalasaṁsthametat |
dhvaniṁ tataḥ saṁchalanādanaṅgakrīḍārasollāsakaraṁ
karoti ||92 ||

अस्मिञ्शय्यातलविनिहिते मुंचति व्यक्तरागं चित्राञ् शब्दान् मृगशिशुदृशां
यान्ति(ति) भीत्येव मानः ।
किंचैतासां दयितमभितो निर्भरप्रेमभाजां ।
प्रौढिं गच्छन्त्यधिकमधिकं मन्मथक्रीडितानि ||93 ||

asmiñ śayyātalavinihite muṁchati vyaktarāgaṁ chitrāñ
śabdān mṛgaśiśudṛśāṁ yānti(ti) bhītyeva mānaḥ |
kiṁchaitāsāṁ dayitamabhito nirbharapremabhājāṁ |
prauḍhiṁ gachchhantyadhikamadhikaṁ
manmathakrīḍitāni ||93 ||

पटहमुरजे वेणुः शङ्खो विपंच्यथ
काहला डमरुटिविले वाद्यातोद्यान्यमून्यखिलान्यपि ।
मधुरमधिकं यच्चित्रं च ध्वनि विदधात्यलं
तदिर विधिना रुद्धोन्मुक्तानिलस्य विजृम्भितम् ||94 ||

paṭahamuraje veṇuḥ śaṅkho vipaṁchyatha
kāhalā ḍamaruṭivile vādyātodyānyamūnyakhilānyapi |
madhuramadhikaṁ yachchitraṁ cha dhvani vidadhātyalaṁ
tadira vidhinā ruddhonmuktānilasya vijṛmbhitam ||94 ||

लघुदारुमयं महाविहङ्गं दृढसुश्लिष्टतनुं विधाय तस्य ।
उदरे रसयन्त्रमादधीत ज्वलनाधारमधेऽस्य चाति(ग्नि)पूर्णम् ||95 ||

laghudārumayaṁ mahāvihaṅgaṁ dṛḍhasuśliṣṭatanuṁ
vidhāya tasya |
udare rasayantramādadhīta jvalanādhāramadhe'sya
chāti(gni)pūrṇam ||95 ||

तत्रारूढः पूरुषस्तस्य पक्षद्वन्द्वोच्चालप्रोज्झितेनानिलेन ।
सुसस्वान्तः पारदस्यास्य शक्त्या चित्रं कुर्वन्नम्बरे याति दूरम् ||96 ||
tatrārūḍhaḥ pūruṣastasya

pakṣadvandvochchālaprojjhitenānilena |
suptasvāntaḥ pāradasyāsya śaktyā chitraṁ kurvannmbare
yāti dūram ||96||

इत्थमेव सुरमन्दिरतुल्यं संचलत्यलघु दारुविमानम् ।
आदधीत विधिना चतुरोऽन्तस्तस्य पारदभृतान् दृढकुम्भान् ॥97॥

itthameva suramandiratulyaṁ saṁchalatyalaghu
dāruvimānam |
ādadhīta vidhinā chaturo'ntastasya pāradabhṛtān
dṛḍhakumbhān ||97||

अयःकपालाहितमन्दवह्नप्रतप्ततत्कुम्भभुवा गुणेन ।
व्योम्नो झगित्याभरणत्वमेति सन्तप्तगर्जद्रसराजशक्त्या ॥98॥

ayaḥkapālāhitamandavahnaprataptatatkumbhabhuvā
guṇena |
vyomno jhagityābharaṇatvameti
santaptagarjadrasarājaśaktyā ||98||

वृत्सन्धितमथायस्यन्त्रं तद् विधाय रसपूरितमन्तः ।
उच्चदेशविनिधापिततप्तं सिंहनादमुरजं विदधाति ॥99॥

vṛttsandhitamathāyasyantraṁ tad vidhāya
rasapūritamantaḥ |
uchchadeśavinidhāpitataptaṁ siṁhanādamurajaṁ
vidadhāti ||99||

स कोऽप्यस्य स्फारः स्फुरति नरसिंहस्य
महिमा पुरस्ताद् यस्यैता मदजलमुचोऽपि द्विपघटाः ।
मुहुः श्रुत्वा निनदमपि गम्भीरविषमं
पलायन्ते भीतास्त्वरितमवधूयाङ्कुशमपि ॥100॥

sa ko'pyasya sphāraḥ sphṛrati narasiṁhasya
mahimā purastād yasyaitā madajalamucho'pi dvipaghaṭāḥ |
muhuḥ śrutvā ninadamapi gambhīraviṣamaṁ
palāyante bhītāstvaritamavadhūyāṅkuśamapi ||100||

दग्ग्रीवातलहस्तप्रकोष्ठबाहूरुहस्तशाखादि ।

सच्छिद्रं वपुरखिलं तत्सन्धिषु खण्डशो घटयेत् ॥101॥

dṛggrīvātalahastaprakoṣṭhabāhūruhastaśākhādi |
sachchhidraṁ vapurakhilaṁ tatsandhiṣu khaṇḍaśo
ghaṭayet ॥101॥

श्लिष्टं कीलकविधिना दारुमयं सृष्टचर्मणा गुप्तम् ।
पुंसोऽथवा युवत्या रूपं कृत्वातिरमणीयम् ॥102॥

śliṣṭaṁ kīlakavidhinā dārumayaṁ sṛṣṭacharmaṇā guptam |
puṁso'thavā yuvatyā rūpaṁ kṛtvātiramaṇīyam ॥102॥

रन्ध्रगतैः प्रत्यङ्ग विधिना नाराचसङ्गतैः सूत्रैः ।
ग्रीवाचलनप्रसरणविकुंचनादीनि विदधाति ॥103॥

randhragataiḥ pratyaṅga vidhinā nārāchasaṅgataiḥ sūtraiḥ |
grīvāchalanaprasaraṇavikuṁchanādīni vidadhāti ॥103॥

करग्रहणताम्बूलप्रदानजलसेचनप्रकाणा(णामा)दि ।
आदर्शप्रतिलोकनवीणावाद्यदि च करोति ॥104॥

karagrahaṇatāmbūlapradānajalasechanaprakāṇā(ṇāmā)di |
ādarśapratilokanavīṇāvādyadi cha karoti ॥104॥

एवमन्यदपि चेदृशमेतत् कर्म विस्मयविधायि विधत्ते ।
जृम्भितेन विधिना निजबुद्धेः कृष्टमुक्तगुणचक्रवशेन ॥105॥

ēvamanyadapi chedṛśametat karma vismayavidhāyi
vidhatte |
jṛmbhitena vidhinā nijabuddheḥ
kṛṣṭamuktaguṇachakravaśena ॥105॥

पुंसो दारुजमूर्ध्वं रूपं कृत्वा निकेतनद्वारी ।
तत्करयोजितदण्डं निरुणद्धि प्रविशतां वर्त्म ॥106॥

puṁso dārujamūrdhvaṁ rūpaṁ kṛtvā niketanadvārī |
tatkarayojitadaṇḍaṁ niruṇaddhi praviśatāṁ vartma ॥106॥

खड्गहस्तमथ मुद्गहस्तं कुन्तहस्तमथवा यदि तत् स्यात् ।
तन्निहन्ति विशतो निशि चौरान् द्वारि संवृतमुखं प्रसभेन ॥107॥

khaḍgahastamatha mudgrahastaṁ kuntahastamathavā yadi

tat syāt l
tannihanti viśato niśi chaurān dvāri saṁvṛtamukhaṁ
prasabhena II107 II

ये चापाद्या ये शतघ्न्यादयोऽस्मि-
न्नुष्ट्रग्रीवाद्याश्च दुर्गस्य गुप्त्यै ।
ये क्रीडाद्याः क्रीडनार्थं च राज्ञां
सर्वेऽपि स्युर्योगतस्ते गुणानाम् ॥108 ॥

ye chāpādyā ye śataghnyādayo'smi-
nnuṣṭragrīvādyāścha durgasya guptyai l
ye krīḍādyāḥ krīḍanārthaṁ cha rājñāṁ
sarve'pi syuryogataste guṇānām II108 II

इदानीं प्रक्रमायातं वरियन्त्रं प्रचक्ष्महे ।
क्रीडार्थं कार्यसिद्ध्यै च चतुर्धा तद्गतिं विदुः ॥109 ॥

idānīṁ prakramāyātaṁ variyantraṁ prachakṣamahe l
krīḍārthaṁ kāryasiddhyai cha chaturdhā tadgatiṁ
viduḥ II109 II

निस्नगं भवति द्रोणीदेशादूर्ध्वस्थिताज्जलम् ।
यत्र तत् पातयन्त्रः स्याद् वाटिकादिप्रयोजनम् ॥110 ॥

nisnagaṁ bhavati droṇīdeśādūrdhvasthitājjalam l
yatra tat pātayantraḥ syād vāṭikādiprayojanam II110 II

उच्छ्रायसमपाताख्यं यत्रोर्ध्वा नाडिका पयः ।
जलाधारगुणान्मुंचेदधस्तात् समनाडिका(कम्) ॥111 ॥

uchchhrāyasamapātākhyaṁ yatrordhvā nāḍikā payaḥ l
jalādhāraguṇānmuṁchedadhastāt samanāḍikā(kam) II111 II

यत्र पातसमुच्छ्रायं पतित्वोच्छ्रायतो जलम् ।
तिर्यग् गत्वा पर्यात्यूर्ध्वं सच्छिद्रस्तम्भयोगतः ॥112 ॥

yatra pātasamuchchhrāyaṁ patitvochchhrāyato jalam l
tiryag gatvā paryātyūrdhvaṁ
sachchhidrastambhayogataḥ II112 II

पतित्वोच्छ्रायतस्तोयं तिर्यगूर्ध्वोर्ध्वमेत्यथ ।
सच्छिद्रस्तम्भयोगेन तत् स्यात् पातसमोच्छ्रयम् ॥113॥

patitvochchhrāyatastoyaṁ tiryagūrdhvordhvametyatha |
sachchhidrastambhayogena tat syāt
pātasamochchhrayam ॥113॥

वाप्यां वापि च कूपे विधानतो दीर्घिकादिका विहिता ।
यत्रोर्ध्वमम्बु गमयति तदिहोच्छ्रयसंज्ञितं कथितम् ॥114॥

vāpyāṁ vāpi cha kūpe vidhānato dīrghikādikā vihitā |
yatrordhvamambu gamayati tadihochchhrayasaṁjñitaṁ
kathitam ॥114॥

दारुजमिभस्य रूपं यत् सलिलं पात्रसंस्थितं पिबति ।
तन्माहात्म्यं निगदितमेतस्योच्छ्रायतुल्यस्य ॥115॥

dārujamibhasya rūpaṁ yat salilaṁ pātrasaṁsthitaṁ pibati |
tanmāhātmyaṁ nigaditametasyochchhrāyatulyasya ॥115॥

सलिलं सुरङ्गदेशानीतं निम्नेन वर्त्मना दूरे ।
अद्भुतमम्भस्थानं तदिह समोच्छ्रायतः कुरुते ॥116॥

salilaṁ suraṅgadeśānītaṁ nimnena vartmanā dūre |
adbhutamambhasthānaṁ tadiha samochchhrāyataḥ
kurute ॥116॥

धारागृहमेकं स्यात् प्रवर्षणाख्यं ततो द्वितीयं च ।
प्राणालं जलमग्नं नन्द्यावर्तं तथान्यदिप ॥117॥

dhārāgṛhamekaṁ syāt pravarṣaṇākhyaṁ tato dvitīyaṁ cha |
prāṇālaṁ jalamagnaṁ nandyāvartaṁ tathānyadipa ॥117॥

प्राकृतजनार्थमेतन्न विधेयं योग्यमेतदवनिभुजाम् ॥
मङ्गल्यानां सदनं दिव्यमिदं तुष्टिपुष्टकरम् ॥118॥

prākṛtajanārthametanna vidheyaṁ
yogyametadavanibhujām ॥
maṅgalyānāṁ sadanaṁ divyamidaṁ tuṣṭipuṣṭakaram ॥118॥

सलिलाशयस्य सविधे कस्याप्याश्रित्य शोभनं देशम् ।

यन्त्रोत्सेधाद् द्विगुणा त्रिगुणा वा नाडिका कार्या ॥119॥

salilāśayasya savidhe kasyāpyāśritya śobhanaṁ deśam l
yantrotsedhād dviguṇā triguṇā vā nāḍikā kāryā ॥119॥

जलनिर्वाहसहासावन्तर्मसृणा बहिश्च नीरन्ध्रा ।
निर्व्यूढाम्भसि तस्यां शुभे मुहूर्ते गृहं कार्यम् ॥120॥

jalanirvāhasahāsāvantarmasṛṇā bahiścha nīrandhrā l
nirvyūḍhāmbhasi tasyāṁ śubhe muhūrte gṛhaṁ
kāryam ॥120॥

सर्वाभिरोषधीभिर्युक्तं सहिरण्यपूर्णकम्भैश्च ।
सुविचित्रगन्धमाल्यं विनादितं ब्रह्मघोषेण ॥121॥

sarvābhiroṣadhībhiryuktaṁ sahiraṇyapūrṇakambhaiścha l
suvichitragandhamālyaṁ vināditaṁ brahmaghoṣeṇa ॥121॥

रत्नोद्भवैर्विचित्रैः स्तम्भैर्युक्तं हिरण्यघटितैर्वा ।
रजतोद्भवैः कदाचित् सुरदारुमसुद्भवैरथवा ॥122॥

ratnodbhavairvichitraiḥ stambhairyuktaṁ
hiraṇyaghaṭitairvā l
rajatodbhavaiḥ kadāchit suradārumasudbhavairathavā ॥122॥

श्रीखण्डोत्थैरथवा सालकमुख्यप्रशास्तवृक्षोत्थैः ।
शतसङ्ख्यैर्द्वात्रिंशत्सङ्ख्यैर्यदि वापि षोडशभिः ॥123॥

śrīkhaṇḍotthairathavā sālakamukhyapraśāstavṛkṣotthaiḥ l
śatasaṅkhyairdvātriṁśatsaṅkhyairyadi vāpi
ṣoḍaśabhiḥ ॥123॥

अथवा चतुस्समन्वितविंशतिसङ्ख्यैर्दिनेशसङ्ख्यैश्चैर्वा ।
भूषितमतिरमणीयैश्चतुर्भिरपि वा विधातव्यम् ॥124॥

athavā chatussamanvitaviṁśatisaṅkhyairdineśasaṅkhyairvā l
bhūṣitamatiramaṇīyaiśchaturbhirapi vā vidhātavyam ॥124॥

प्राग्ग्रीवैरतिचित्रैः शलैर्जलैर्विभूषितं विविधैः ।
वेदीभिः परिकरितं कपोतपालीभिरभिरामम् ॥125॥

prāggrīvairatichitraiḥ śalairjālairvibhūṣitaṁ vividhaiḥ l

vedībhiḥ parikaritaṁ kapotapālībhirabhirāmam ॥125॥

रमणीयसालभंजिकमनेकविधयन्त्रशकुनिकृतशोभम् ।
मिथुनैश्च वनराणां जम्भकनिवहैश्च नैकविधैः ॥126॥

*ramaṇīyasālabhaṁjikamanekavidhayantraśakunikṛtaśobham
mithunaiścha vanarāṇāṁ jambhakanivahaiścha
naikavidhaiḥ ॥126॥*

विद्याधरसिद्धभुजङ्गकिन्नरैश्चारणैश्च रमणीयम् ।
नृत्यद्भिः परमग(ग)णैः शिखण्डिभिर्मण्डितोद्देशम् ॥127॥

*vidyādharasiddhabhujaṅgakinnaraiśchāraṇaiścha
ramaṇīyam /
nṛtyadbhiḥ paramaga(ga)ṇaiḥ
śikhaṇḍibhirmaṇḍitoddeśam ॥127॥*

कल्पतरुभिर्विचत्रैश्चित्रलतावल्लिगुल्संछन्नम् ।
परपुष्टषद्दालीमरालमालामनोहारि ॥128॥

*kalpatarubhirvichatraiśchitralatāvalligulsaṁchhannam /
parapuṣṭaṣaṭpadālīmarālamālāmanohāri ॥128॥*

प्रवहत्सकलस्रोतःसुश्लिनिविष्टनाडिकं मध्ये ।
सच्छिद्रनाडिकयुतं नानाविधरूपरमणीयम् ॥129॥

*pravahatsakalasrotaḥsuśliniviṣṭanāḍikaṁ madhye /
sachchhidranāḍikayutaṁ nānāvidharūparamaṇīyam ॥129॥*

सुश्लिष्टनाडिकाग्रे स्तम्भतुलाभित्तिसंश्रिते परितः ।
सम्यक् कृत्वा दृढतरविलेपनं वज्रलेपाद्यैः ॥130॥

*suśliṣṭanāḍikāgre stambhatulābhittisaṁśrite paritaḥ /
samyak kṛtvā dṛḍhataravilepanaṁ vajralepādyaiḥ ॥130॥*

लाक्षासरर्जरसद्दृषन्मेषविषाणोत्थचूर्णसंमिश्रम् ।
अतसीकरंजतैलप्रविगाढो वज्रलेपः स्यात् ॥131॥

*lākṣāsararjarasadṛṣanmeṣaviṣāṇotthachūrṇasaṁmiśram /
atasīkaraṁjatailapravigāḍho vajralepaḥ syāt ॥131॥*

दृढसन्धिबन्धहेताः स तत्र देयो द्विशः कदाचिद् वा ।

शणवल्कश्लेष्मातकसिक्थकतैलैः प्रलेपश्च ॥132 ॥

dṛḍhasandhibandhahetāḥ sa tatra deyo dviśaḥ kadāchid vā |
śaṇavalkaśleṣmātakasikthakatailaiḥ pralepaścha ॥132 ॥

उच्छ्रययन्त्रेणेतद् भ्रान्तजलेनाथ तदभितः कृत्वा।
चित्रानुपातयुक्तं प्रदर्शयेन्नृपतये स्थपतिः ॥133 ॥

uchchhrayayantreṇetad bhrāntajalenātha tadabhitaḥ kṛtvā |
chitrānupātayuktaṁ pradarśayennṛpataye sthapatiḥ ॥133 ॥

कार्याण्यस्मिन् करिणां मिथुनान्यभितोऽम्बुकेलियुक्तानि ।
अन्योन्यपुष्कारोंझ्रतसीकरभयपिहितनयनानि ॥134 ॥

kāryāṇyasmin kariṇāṁ mithunānyabhito'mbukeliyuktāni |
anyonyapuṣkāroṁjhratasīkarabhayapihitanayanāni ॥134 ॥

वर्षानुकृतं चास्मिन् प्रीतिमति प्रतिमङ्गजो वीक्ष्य ।
दृक्कटमेहनहस्तैर्मदमिव मुंचञ् जलं कार्यः ॥135 ॥

varṣānukṛtaṁ chāsmin prītimati pratimaṅgajo vīkṣaya |
dṛkkaṭamehanahastairmadamiva muṁchañ jalaṁ
kāryaḥ ॥135 ॥

स्तनयोर्युगेन सृजी जलधारे तत्र कापि कार्या स्त्री ।
आनन्दाश्रुलवानिव सलिलकणान् पक्ष्मभिः काचित् ॥136 ॥

stanayoryugena sṛjī jaladhāre tatra kāpi kāryā strī |
ānandāśrulavāniva salilakaṇān pakṣamabhiḥ kāchit ॥136 ॥

नाभिह्रदनदिकामिव विनिर्गतां कापि बिभ्रती धाराम् ।
काप्यङ्गलीनखांशुभिरिव योषित् सिंचती कार्या ॥137 ॥

nābhihradanadikāmiva vinirgatāṁ kāpi bibhratī dhārām |
kāpyaṅgalīnakhāṁśubhiriva yoṣit siṁchatī kāryā ॥137 ॥

एवम्प्रायांश्चित्रान् स्वभावचेष्टान् बहूंश्च रमणीयान् ।
क्षोभान् विधाय कुर्यादाश्चर्यं नरपतेः स्थपतिः ॥138 ॥

ēvamprāyāṁśchitrān svabhāvacheṣṭān bahūṁścha
ramaṇīyān |
kṣobhān vidhāya kuryādāścharyaṁ narapateḥ

sthapatiḥ ॥138॥

मध्ये तस्य विधेयं सिंहासनममलहेममणिघटितम् ।
तत्रासीदेन्नरपतिरवनिपतिः श्रीपतिर्देवः ॥139॥

*madhye tasya vidheyaṁ
siṁhāsanamamalahemamaṇighaṭitam /
tatrāsīdennarapatiravanipatiḥ śrīpatirdevaḥ* ॥139॥

स्नायात् कदाचिदस्मिन् मङ्गलगीतैर्विवर्धितानन्दः ।
वादित्रनदात्यनिपुणैर्निषेव्यमाणः सुरेन्द्र इव ॥140॥

*snāyāt kadāchidasmin maṅgalagītairvivardhitānandaḥ /
vāditranadātyanipuṇairniṣevyamāṇaḥ surendra iva* ॥140॥

य एतस्मिन् गाढग्लपितघनघर्मव्यतिकरे
शुचौ धाराधाम्नि स्फुटसलिलधारे नरपतिः ।
सुखेना से पश्यन् विविधजलशिल्पानि स
भवेन्न मर्त्यः किन्त्वेष क्षितिकृतनिवासः सुरपतिः ॥141॥

*ya ētasmin gāḍhaglapitaghanagharmavyatikare
śuchau dhārādhāmni sphuṭasaliladhāre narapatiḥ /
sukhenāse paśyan vividhajalaśilpāni sa
bhavenna martyaḥ kintveṣa kṣitikṛtanivāsaḥ surapatiḥ* ॥141॥

जलदकुलाष्टकयुक्तं पूर्ववदन्यद् गृहं समारचयेत् ।
वर्षद्वारानिकरैः पर्वर्षणाख्यां तदाप्नोति ॥142॥

*jaladakulāṣṭakayuktaṁ pūrvavadanyad gṛhaṁ
samārachayet /
varṣadvārānikaraiḥ parvarṣaṇākhyāṁ tadāpnoti* ॥142॥

प्रतिकुलमस्मिन् कार्या दिव्यालङ्कारधारिणः पुरुषाः ।
विधिना त्रयः सुरूपाश्चत्वारः सप्त वा सुदृढाः ॥143॥

*pratikulamasmin kāryā divyālaṅkāradhāriṇaḥ puruṣāḥ /
vidhinā trayaḥ surūpāśchatvāraḥ sapta vā sudṛḍhāḥ* ॥143॥

यन्त्रेण समोच्छ्रायेण तांश्चतुर्थेन वा ततः पुरुषान् ।
कृत्वा सवक्रनालानम्भोभिः पूरयेद् विमलैः ॥144॥

yantreṇa samochchhrāyeṇa tāṁśchaturthena vā tataḥ puruṣān |
kṛtvā savakranālānambhobhiḥ pūrayed vimalaiḥ ||144 ||

सलिलप्रवेशरन्ध्राण्यखिलानि पिधाय तत्र पुरुषाणाम् ।
अङ्गानि वारिमोक्षाण्यखिलन्यथ मोचयेत् तेषाम् ॥145॥

salilapraveśarandhrāṇyakhilāni pidhāya tatra puruṣāṇām |
aṅagāni vārimokṣāṇyakhilanyatha mochayet teṣām ||145 ||

सलिलं सवक्रनालं द्वारप्रतिरोधमोचनैः पुरुषाः ।
मुंचन्ति स्वेच्छममी विचित्रपातेन चित्रकरम् ॥146॥

salilaṁ savakranālaṁ dvārapratirodhamochanaiḥ puruṣāḥ |
muṁchanti svechchhamamī vichitrapātena chitrakaram ||146 ||

इत्थमिमान् वारिधरान् सामस्या(स्त्या)द्व्यन्तरेण वा सलिलम् ।
त्र्यन्तरतो वा स्वेच्छं प्रवर्षयेदतिमहच्चिन्नम् ॥147॥

itthamimān vāridharān sāmasyā(styā)dvyantareṇa vā salilam |
trayantarato vā svechchhaṁ pravarṣayedatimahachchhinnam ||147 ||

इदं नानाकारं कुलभवनमाद्यं रतिपतेनिर्वास-
श्चित्राणामनुकरणमेकं जलमुचाम् ।
पयःपातैर्ग्रीष्मे रविकरपरीतापशमनं
न केषामत्थर्यं भवति नयनानन्दजननम् ॥148॥

idaṁ nānākāraṁ kulabhavanamādyaṁ ratipatenirvāsa-
śchitrāṇāmanukaraṇamekaṁ jalamuchām |
payaḥpātairgrīṣme ravikaraparītāpaśamanaṁ
na keṣāmattharyaṁ bhavati nayanānandajananam ||148 ||

एकेनाथ चतुर्भिः स्तम्भैरष्टभिरर्थार्किसङ्ख्यैर्वा ।
षोडशभिर्वा कुर्यान्मनोहरं गृहमिह द्वितलम् ॥149॥

ēkenātha chaturbhiḥ
stambhairaṣṭabhirarthārkasaṅkhyairvā |

ṣoḍaśabhirvā kuryānmanoharaṁ gṛhamiha dvitalam ॥149॥

भद्रैर्युतं चतुर्भिश्चतुरश्रं सर्वभित्तिसंयुक्तम् ।
ईलीतोरणयुक्तं कर्तव्यं पुष्पकाकारम् ॥150॥

bhadrairyutaṁ chaturbhischaturaśraṁ
sarvabhittisaṁyuktam ।
īlītoraṇayuktaṁ kartavyaṁ puṣpakākāram ॥150॥

तस्योपरि मध्यगता प्राङ्गणवापी दृढा विधातव्या ।
शतपत्रविहितभूषा तन्मध्ये कर्णिका कार्या ॥151॥

tasyopari madhyagatā prāṅgaṇavāpī dṛḍhā vidhātavyā ।
śatapatravihitabhūṣā tanmadhye karṇikā kāryā ॥151॥

तल्कोणेषु चतुर्ष्वपि रमणीया दारुदारिकाः कार्याः ।
मध्याम्बुजनिहितदृशः सालङ्कारा सश्रृङ्गाराः ॥152॥

tatkoṇeṣu chaturṣvapi ramaṇīyā dārudārikāḥ kāryāḥ ।
madhyāmbujanihitadṛśaḥ sālaṅkārā saśṛṅgārāḥ ॥152॥

पूर्वोक्तयन्त्रयोगात् पद्मसीने वसुन्धराधिपतौ ।
भृङ्गारामलवारिभिरङ्गणवापी भ्रियाच्च ततः ॥153॥

pūrvoktayantrayogāt padmasīne vasundharādhipatau ।
bhṛṅgārāmalavāribhiraṅgaṇavāpī bhriyāchcha tataḥ ॥153॥

तामिति भृत्वा वापीं तत्सलिलं तदनुपट्टगर्भगतम् ।
छाद्यस्तु गन्धरोध्रेष्वति रोहति(?)सर्वतो नियतम् ॥154॥

tāmiti bhṛtvā vāpīṁ tatsalilaṁ tadanupaṭṭagarbhagatam ।
chhādyastu gandharodhreṣvati rohati(?)sarvato
niyatam ॥154॥

मखपट्टसमुत्कीर्णे रूपैश्चित्रैर्मनोरमैरखिलैः ।
अङ्गैर्वारि विमुंचति नासास्यश्रवणनेत्राद्यैः ॥155॥

makhapaṭṭasamutkīrṇai rūpaischitrairmanoramairakhilaiḥ ।
aṅgairvāri vimuṁchati nāsāsyaśravaṇanetrādyaiḥ ॥155॥

प्रणालाख्यं धाराभवनमिदमत्यद्भुततरं स्थितिं
धत्ते यस्य क्षितिपतिलकस्याङ्गणभुवि ।

करोत्येतद् वेत्थं स्थपतिरपि बुद्ध्या
चतुरया जगत्येतौ द्वावप्यधिकमहनीयौ कृतधियाम् ॥156॥

praṇālākhyaṁ dhārābhavanamidamatyadbhutataraṁ sthitiṁ
dhatte yasya kṣitipatilakasyāṅganabhuvi |
karotyetad vetthaṁ sthapatirapi buddhyā
chaturayā jagatyetau dvāvapyadhikamahanīyau
kṛtadhiyām ||156 ||

चतुरश्रातिगभीरा वापी कार्या मनोरमा सुदृढा ।
गर्भगतं गृहमस्याः कर्तव्यं लिप्तसन्धि ततः ॥157॥

chaturaśrātigabhīrā vāpī kāryā manoramā sudṛḍhā |
garbhagataṁ gṛhamasyāḥ kartavyaṁ liptasandhi
tataḥ ||157 ||

विहितप्रवेशनिर्गति सुरङ्ग्याधो निवेशितद्वारम् ।
विदधीत चारुरूपैः प्रवर्षकैर्व्याप्तमुरष्टिात् ॥158॥

vihitapraveśanirgati suraṅgayādho niveśitadvāram |
vidadhīta chārurūpaiḥ pravarṣakairvyaptamuraṣṭiāt ||158 ||

चित्राध्यायोदितवर्त्मना ततोऽलङ्कृतं च चित्रेण ।
तस्य विधेयं मध्यं सलिलधिपवाससङ्काशम् ॥159॥

chitrādhyāyoditavartmanā tato'laṅkṛtaṁ cha chitreṇa |
tasya vidheyaṁ madhyaṁ saliladhipavāsasaṅkāśam ||159 ||

ऊर्ध्वविनिर्गमिताब्जैनालैस्तत्पट्टकन्दकोद्भूतैः ।
सच्छिद्रकर्णिकागतदिनकरकरनिर्मितोद्द्योतम् ॥160॥

ūrdhvavinirgamitābjainālaistatpaṭṭakandakodbhūtaiḥ |
sachchhidrakarṇikāgatadinakarakaranirmitoddyotam ||160 ||

आपूरयेत् ततोऽनु च पाताम्बुभिरमलकमलपर्यन्तम् ।
विधिनामुनैव सम्यक् प्रविधाय मनोरमं भवनम् ॥161॥

āpūrayet tato'nu cha pātāmbubhiramalakamalaparyantam |
vidhināmunaiva samyak pravidhāya manoramaṁ
bhavanam ||161 ||

नानारूपकयुक्त्याउ(व्यु)परचिततमङ्गतोरणद्वारम् ।
शालाभिरायताभिश्चतसृष्वपि दिक्षु कृतशोभम् ॥162 ॥

nānārūpakayuktyāu(vyu)parachitatamaṅgatoraṇadvāram /
śālābhirāyatābhiśchatasṛṣvapi dikṣu kṛtaśobham ॥162 ॥

कृत्रिमशफरीमकारीपक्षिभिरपि चाम्बुसम्भवैर्युक्ताम् ।
कुर्यादम्भोजवतीं वापीमाहार्ययोगेन ॥163 ॥

kṛtrimaśapharīmakārīpakṣibhirapi
chāmbusambhavairyuktām /
kuryādambhojavatīṁ vāpīmāhāryayogena ॥163 ॥

सामन्तमुख्यपुरुषा रारजाज्ञालब्धसंश्रयास्तत्र ।
परराष्ट्रगतदूतास्तिष्ठेयुर्नि हितमिह निभृताः ॥164 ॥

sāmantamukhyapuruṣā rārajājñālabdhasaṁśrayāstatra /
pararāṣṭragatadūtāstiṣṭheyurni hitamiha
nibhṛtāḥ ॥164 ॥

अथ स यथाविधि सलिलक्रीडां पूर्वोक्तमार्गरूपाणाम् ।
दृष्ट्वा मुदितः कुर्यात् पर्यङ्कारोहणं नृपतिः ॥165 ॥

atha sa yathāvidhi salilakrīḍāṁ pūrvoktamārgarūpāṇām /
dṛṣṭavā muditaḥ kuryāt paryaṅkārohaṇaṁ nṛpatiḥ ॥165 ॥

तत्र स्थितस्य नृपतेः परिवारितस्य
वाराङ्गनाभिरभितो जलमग्नधाम्नि ।
पातालसद्मनि यथा भुजगेश्वरस्य
निस्सीमसम्भृतरतिर्भवति प्रमोदः ॥166 ॥

tatra sthitasya nṛpateḥ parivāritasya
vārāṅganābhirabhito jalamagnadhāmni /
pātālasadmani yathā bhujageśvarasya
nissīmasambhṛtaratirbhavati pramodaḥ ॥166 ॥

पूर्वोक्तवापिकायां मध्ये स्तम्भैश्चतुर्भिरुपरचितम् ।
मुक्ताप्रवालयुक्तं पुष्पकमथ कारयेल्लटभम् ॥167 ॥

pūrvoktavāpikāyāṁ madhye
stambhaiśchaturbhirūparachitam /

muktāpravālayuktaṁ puṣpakamatha kārayellaṭabham ||167||

वापीं परितः पुष्पकमापूर्य सुनिर्गमाभिरथ सुदृढम् ।
गभस्वस्तिकभित्तिभिरूपहितशोभं समन्ततः कुर्यात् ॥168॥

vāpīṁ paritaḥ puṣpakamāpūrya sunirgamābhiratha sudṛḍham |
gabhasvastikabhittibhirūpahitaśobhaṁ samantataḥ kuryāt ||168||

पूर्वोक्तवारियोगात् पूर्णमाकर्णतो विधायैताम् ।
जलकेलिषु सोत्क्ण्ठो महीपतिः पुष्पकं यायात् ॥169॥

pūrvoktavāriyogāt pūrṇāmākarṇato vidhāyaitām |
jalakeliṣu sotkṇṭho mahīpatiḥ puṣpakaṁ yāyāt ||169||

कुर्वीत नर्मसचिवैर्विलासिनीभिश्च सार्धमवनिपतिः ।
तद्भित्त्यन्तरवर्ती निमज्जनोन्मज्जनैः क्रीडाम् ॥170॥

kurvīta narmasachivairvilāsinībhiścha sārdhamavanipatiḥ |
tadbhittyantaravartī nimajjanonmmajjanaiḥ krīḍām ||170||

एकत्र मग्नैरपत्र दृष्टैरन्यत्र हत्वा सलिलेन *नष्टैः* ।
क्रीडत्यलं केलिकरः सहायैनृपः सुखं मज्जनपुष्करिण्याम् ॥171॥

ēkatra magnairapatra dṛṣṭairanyatra hatvā salilena nṣṭaiḥ |
krīḍatyalaṁ kelikaraḥ sahāyainṛpaḥ sukhaṁ majjanapuṣkariṇyām ||171||

वापीतलस्थितमथ त्रपयावनम्रमाच्छादितस्तनभरं करपल्लेवन ।
गाढावसक्तवसनं जलरोध्मुक्तावालोकते प्राणयिनीजनमत्र धन्यः ॥172॥

vāpītalasthitamatha trapayāvanamramāchchhāditastanabharaṁ karapallevana |
gāḍhāvasaktavasanaṁ jalarodhmuktāvālokate prāṇayinījanamatra dhanyaḥ ||172||

रथदोलादिविधानं दारवमभिदध्महे वयं सम्यक् ॥
यन्त्रभ्रमणककर्म प्रकीर्तितं पंचमं यत् यत् ॥173॥

rathadolādividhānaṁ dāravamabhidadhmahe vayaṁ

samyak ||
yantrabhramaṇakakarma prakīrtitaṁ paṁchamaṁ yat
yat ||173||

तत्र वसन्तः प्रथमो मदननिवासां वसनततिलकश्च ।
विभ्रमकस्त्रिपुराख्यः पंचेते दोलकाः कथिताः ॥174॥

tatra vasantaḥ prathamo madananivāsāṁ
vasanatatilakaścha |
vibhramakastripurākhyaḥ paṁchete dolakāḥ
kathitāḥ ||174||

खिनेच्चतुरः स्तम्भान् समैकसूत्रोपगान् ऋजून् सुदृढान् ।
सदृशान्तरान् धरित्रीवशतः सुश्लिक्षण(ष्ट)पीठगतान् ॥175॥

khinechchaturaḥ stambhān samaikasūtropagān ṛjūn
sudṛḍhān |
sadṛśāntarān dharitrīvaśataḥ suślikṣaṇa(ṣṭa)pīṭhagatān ||175||

प्रासादस्योक्तदिशि प्रविदध्याद् विरचिताष्टकरदैर्ध्यम् ।
भूमिगृहं रमणीयं तदर्धतो विहितगाम्भीर्यम् ॥176॥

prāsādasyoktadiśi pravidadhyād virachitāṣṭakaradairdhyam |
bhūmigṛham ramaṇīyaṁ tadardhato
vihitagāmbhīryam ||176||

तद्गर्भतले स्तम्भो लोहमयाधारसंस्थितः कार्यः ।
भ्रमसहितः पीठयुतो ग्रस्तश्चच्छादकतुलाभिः ॥177॥

tadgarbhatale stambho lohamayādhārasaṁsthitaḥ kāryaḥ |
bhramasahitaḥ pīṭhayuto
grastaśchachchhādakatulābhiḥ ||177

संस्थाप्योपरि पीठस्य कुम्भिकामतिदृढां विभक्तां च ।
धनुरुच्छ्रितस्ततोऽमूमष्टभिरावेष्टयेद् भद्रैः ॥178॥

samsthāpyopari pīṭhasya kumbhikāmatidṛḍhāṁ vibhaktāṁ
cha |
dhanuruchchhritastato'mūmaṣṭabhirāveṣṭayed
bhadraiḥ ||178||

स्वेच्छमथ भूमिकोच्छ्रयमस्योर्ध्वे कल्पयेन्नितान्तमृजुम् ।
निदधीत वेष्टनोर्ध्वे पट्टयुतं स्तम्भशीर्षं च ॥179॥

*svechchhamatha bhūmikochchhrayamasyordhve
kalpayennitāntamṛjum |
nidadhīta veṣṭanordhve paṭṭayutaṁ stambhaśīrṣaṁ
cha ॥179॥*

हीरग्रह(ण?)पर्यन्तं मदला गजशीर्षिका विधातव्या ।
सुदृढा प्रयत्नरचिता मनोभिरामा यथाशोभम् ॥180॥

*hīragraha(ṇa?)paryantaṁ madalā gajaśīrṣikā vidhātavyā |
sudṛḍhā prayatnarachitā manobhirāmā yathāśobham ॥180॥*

पट्टस्योपरि कार्या चतुष्किका क्षेत्रमानतोऽभीष्टात् ।
तस्यामुपरि विधेयस्तलबन्धो दृढतरन्यासः ॥181॥

*paṭṭasyopari kāryā chatuṣkikā kṣetramānato'bhīṣṭāt |
tasyāmupari vidheyastalabandho dṛḍhataranyāsaḥ ॥181॥*

स्तम्भैर्द्वादशभिरथ क्षेत्रे युक्त्या समुच्छ्रितैर्भव्यैः ।
रूपवतीकोणस्थितिराधिका भूः प्रथमिका कार्या ॥182॥

*stambhairdvādaśabhiratha kṣetre yuktyā
samuchchhritairbhavyaiḥ |
rūpavatīkoṇasthitirādhikā bhūḥ prathamikā kāryā ॥182॥*

मध्ये भ्रमश्च तस्या गर्भस्तम्भप्रतिष्ठितः कार्यः ।
क्षेत्रप्रमाणवशतस्तां पश्चाच्छादयेत् पट्टैः ॥183॥

*madhye bhramaścha tasyā garbhastambhapratiṣṭhitaḥ
kāryaḥ |
kṣetrapramāṇavaśatastāṁ paśchāchchhādayet paṭṭaiḥ ॥183॥*

रथिकाशिखाग्रकेषु च फलकाम(व) रणस्य तद्वदुपरिष्टात् ।
भ्रमचक्राणि न्यस्येन्मध्ये स्तम्भे च पंचैव ॥184॥

*rathikāśikhāgrakeṣu cha phalakāma(va) raṇasya
tadvadupariṣṭāt |
bhramachakrāṇi nyasyenmadhye stambhe cha
paṁchaiva ॥184॥*

अत उपरि यथाशोभं हि भूमिका पुष्पकाकृतिः कार्या।
मध्यस्तम्भाधारा कृतकलशविभूष्णा शिरसि ॥185॥

ata upari yathāśobhaṁ hi bhūmikā puṣpakākṛtiḥ kāryā |
madhyastambhādhārā kṛtakalaśavibhūṣṇā śirasi ||185||

स्तम्भेऽव(ध)स्ताद् भ्रमिते भृशं भ्रमत्यर्थभूमिका तत्र।
रथकाभ्रमरकयुक्ता परस्परं चक्रयन्त्रेण ॥186॥

stambhe'va(dha)stād bhramite bhṛśaṁ
bhramatyarthabhūmikā tatra |
rathakābhramarakayuktā parasparaṁ chakrayantreṇa ||186||

वसन्तरथिकाभ्रमे समधिरूढवाराङ्गना
परिभ्रमणसम्भृताभ्यधिकविभ्रमं भूपतिः।
करोति नयनोत्सवस्त्रि(वं त्रि) दशधाम्नि
यत्कीर्तनं वसन्तसमये भवत्यमलकीर्त्तिधामैव सः ॥187॥

vasantarathikābhrame samadhirūḍhavārāṅganā
paribhramaṇasambhṛtābhyadhikavibhramaṁ bhūpatiḥ |
karoti nayanotsavastri(vaṁ tri) daśadhāmni
yatkīrtanaṁ vasantasamaye bhavatyamalakīrttidhāmaiva
saḥ ||187||

आरोप्य स्थिरमेकं स्तम्भं भूमिगृहादिरहितमथ।
हस्तचतुष्कोच्छ्राया कार्योपरि भूमिका चास्य ॥188॥

āropya sthiramekaṁ stambhaṁ bhūmīgṛhādirahitamatha |
hastachatuṣkochchhrāyā kāryopari bhūmikā chāsya ||188||

मध्ये भ्रमरकयुक्तं शेषं पूर्ववदिहाचरेदखिलम्।
पुष्पकमिपि च स्तम्भे शिथिलं कलशोच्छ्रितं कुर्यात् ॥189॥

madhye bhramarakayuktaṁ śeṣaṁ
pūrvavadihācharedakhilam |
puṣpakamipi cha stambhe śithilaṁ kalaśochchhritaṁ
kuryāt || 189 ||

तस्योपरि च ग्रीवा चतुरासनसंयुता विधातव्या।
घण्टास्तम्भौ कार्यौ स्तम्भेन महाबलौ तत्र ॥190॥

tasyopari cha grīvā chaturāsanasaṁyutā vidhātavyā |
ghaṇṭāstambhau kāryau stambhena mahābalau tatra ||190 ||

एवं पुष्पकभूमिकान्तरलस्थायी निगूढो
जनो यावद् भ्रामकयन्त्रचक्रनिकरं सम्यक् क्रमाच्चलयेत् ।
तावत् ता रथिकासना मृगदृशस्तत्र स्थिताः
पुष्पके कामावासकुतूहलार्पितदृशो भ्राम्यन्ति सर्वा अपि ||191 ||

ēvaṁ puṣpakabhūmikāntaralasthāyī nigūḍho
jano yāvad bhrāmakayantrachakranikaraṁ samyak
kramāchchalayet |
tāvat tā rathikāsanā mṛgadṛśastatra sthitāḥ
puṣpake kāmāvāsakutūhalārpitadṛśāo bhrāmyanti sarvā
api ||191 ||

अथ कोणगतान् स्तम्भांश्चतुरो विनिवेशयद क्रजून् सुदृढान् ।
सुश्लिष्टपीठसंस्थान् समान्तरान् मेदिनीवशतः ||192 || ।

atha koṇagatān stambhāṁśchaturo viniveśayada krajūn
sudṛḍhān |
suśliṣṭapīṭhasaṁsthān samāntarān medinīvaśataḥ ||192 ||

तेषामुपरि लता(तला) न्तरसंयुक्ता भूमिका विधातव्या ।
रथिकास्तत्र चतस्रो जायन्ते पूर्ववद् दिक्स्थाः ||193 ||

teṣāmupari latā(talā) ntarasaṁyuktā bhūmikā vidhātavyā |
rathikāstatra chatasro jāyante pūrvavad diksthāḥ ||193 ||

तदुपरि तथार्धभूमिः कार्या सुश्लिष्टदारुसन्धाना ।
मध्यभ्ररकयुक्ता सरूपका मत्तवारणयुता च ||194 ||

tadupari tathārdhabhūmiḥ kāryā suśliṣṭadārusandhānā |
madhyabhrarakayuktā sarūpakā mattavāraṇayutā cha ||194 ||

नानाविधकर्मवती वसन्तो बाह्यरेखा स्यात् ।
अन्योन्ययन्त्रपरिघट्टनदोल्यमान-
निश्शेषचक्ररथिकाभ्रमणाभिरारमम् ।
दृष्ट्वा वसन्ततिलकं सुरमन्दिराणां

भूषायमाणमुपयाति न विस्मयत्वं(यं कः) ॥195॥

nānāvidhakarmavatī vasanto bāhyarekhā syāt |
anyonyayantraparighaṭṭanadolyamāna-
niśśeṣachakrarathikābhramaṇābhirāramam |
dṛṣṭavā vasantatilakaṁ suramandirāṇāṁ
bhūṣāyamāṇamupayāti na vismayatvaṁ(yaṁ kaḥ) ॥195॥

प्रविधाय रङ्गभूमिं प्रथमां शास्त्रान्तराधरस्यार्थे(?)
चतुरश्रा रूपवती सचतुर्भद्रा विधेया भूः ॥196॥

pravidhāya raṅgabhūmiṁ prathamāṁ
śāstrāntarādharasyārthe(?)
chaturaśrā rūpavatī sachaturbhadrā vidheyā bhūḥ ॥196॥

प्रतिकोणमागता(ता) स्या भद्रेषु भवन्ति संयता भ्रमराः ।
अत उपरिष्टाद् भूम्या भ्रमराश्चाष्टासनाः कार्याः ॥197॥

pratikoṇamāgatā(tā) syā bhadreṣu bhavanti saṁyatā
bhramarāḥ |
ata upariṣṭād bhūmyā bhramarāśchāṣṭāsanāḥ kāryāḥ ॥197॥

रेखाः शुद्धाः कार्या बहिरन्तश्चित्रिताश्चान्याः ।
पीठेषु मध्यग(सं) स्थास्ततोऽपरा भूमिकाः कार्याः ॥198॥

rekhāḥ śuddhāḥ kāryā bahirantaśchitritāśchānyāḥ |
pīṭheṣu madhyaga(saṁ) sthāstato'parā bhūmikāḥ
kāryāḥ ॥198॥

पीठस्य मध्यसंस्थैरन्योन्याराऴियोजितैश्चक्रैः ।
सर्वे वेगाद् भ्राम्यन्ति सान्तन(रा) विभ्रमे भ्रमराः ॥199॥

pīṭhasya madhyasaṁsthairanyonyārāliyojitaiśchakraiḥ |
sarve vegād bhrāmyanti sāntana(rā) vibhrame
bhramarāraḥ ॥199॥

दोलासनो विहितवारवधूकृ(भृ)ता-
तिचित्रेण यत्त्रिदशधामसु विभ्रमेण ।
पृथ्वीपतिमुदमुपैति समुल्लसन्ती
कीर्त्तिर्न माति भुवनत्रितयेऽपि तस्य ॥200॥

dolāsano vihitavāravadhūkṛ(bhṛ)tā-
tichitreṇa yastridaśadhāmasu vibhrameṇa |
pṛthvīpatimudamupaiti samullasantī
kīrttirna māti bhuvanatritaye'pi tasya ||200||

चतुरश्रमथ क्षेत्रं कृत्वांशैर्भाजितं ततोऽष्टाभिः ।
काणैः शेषैस्तसिंमश्चतुरश्रं कल्पयेद् भद्रम् ॥201॥

chaturaśramatha kṣetraṁ kṛtvāṁśairbhājitaṁ tato'ṣṭabhiḥ |
kāṇaiḥ śeṣaistasimmaśchaturaśraṁ kalpayed bhadram ||201||

तद्द्विगुणमूर्ध्वमेतस्य भूमिकाभागसङ्ख्यया कार्यम् ।
तत्राद्यंशचतुष्केण भूमिका स्यात् समुच्छ्रयतः ॥202॥

taddviguṇamūrdhvametasya bhūmikābhāgasaṅkhyayā
kāryam |
tatrādyaṁśachatuṣkeṇa bhūmikā syāt
samuchchhrayataḥ ||202||

तत्राष्टषड्तुर्भागवर्जिता भूमिका उपर्युपरि ।
क्रमशो भवन्त्यथैवं ताः स्युस्तिस्रोऽर्धसंयुक्ताः ॥203॥

tatrāṣṭaṣaṭchaturbhāgavarjitā bhūmikā uparyupari |
kramaśo bhavantyathaivaṁ tāḥ
syustisro'rdhasaṁyuktāḥ ||203||

शेषांशोच्छ्रययुक्ता घण्टा चतुरश्रकायता कार्या ।
त्रिचतुर्भूम्यौ कार्ये सषड्तुर्भागविस्तारे ॥204॥

śeṣāṁśochchhrayayuktā ghaṇṭā chaturaśrakāyatā kāryā |
trichaturbhūmyau kārye saṣaṭchaturbhāgavistāre ||204||

रङ्गस्यादाद्यभूवि द्वितीयभुवि कोणगास्तथा रथिकाः ॥
स्युर्भटकृतियुक्ता दोला अपि तत्र रमणीयाः ॥205॥

raṅgasyādādyabhūvi dvitīyabhuvi koṇagāstathā rathikāḥ ||
syurbhadṛkṛtiyuktā dolā api tatra ramaṇīyāḥ ||205||

रथिकास्तृतीयभूमौ कार्या भद्रेषु चातिरमणीयाः ।
कोणेष्वथासनान्यर्ध्वास्तुकेऽपि भ्रमः कार्यः ॥206॥

rathikāstṛtīyabhūmau kāryā bhadreṣu chātiramaṇīyāḥ |
koṇeṣvathāsanānyardhavāstuke'pi bhramaḥ kāryaḥ ||206 ||

दोलारथिक चतुरासने भ्रमोष्टासनो भवेत् तत्र ।
आसनमिह तत् कथितं युवतेः स्थानं यदेकं स्यात् ॥207 ॥

dolārathika chaturāsane bhramoṣṭāsano bhavet tatra |
āsanamiha tat kathitaṁ yuvateḥ sthānaṁ yadekaṁ
syāt ||207 ||

निखिलान्यपि भ्रमणसंमुखं तानि बिभ्रति भ्रमणम्(?)
यत्रासनानि स इह भ्रम इत्युक्तोऽपराधिका(?) ॥208 ॥

nikhilānyapi bhramaṇasaṁmukhaṁ tāni bibhrati
bhramaṇam(?)
yatrāsanāni sa iha bhrama ityukto'parādhikā(?) ||208 ||

यष्टेरूर्ध्वमधस्ताद् भ्रमस्य चक्रं(नि) योजयेदेकम् ।
लघुचक्राणि च तद्वन्नियोजयेदासनेष्वत्र ॥209 ॥

yaṣṭerūrdhvamadhastād bhramasya chakraṁ(ni)
yojayedekam |
laghuchakrāṇi cha tadvanniyojayedāsaneṣvatra ||209 ||

लघुचक्रारकवृत्ते संलग्नाः कीलका दृढाः कार्याः ।
तुल्यान्तराः समस्ताः प्रलघु(क) चक्रारवृन्तगताः ॥210 ॥

laghuchakrārakavṛtte saṁlagnāḥ kīlakā dṛḍhāḥ kāryāḥ |
tulyāntarāḥ samastāḥ pralaghu(ka)
chakrāravṛntagatāḥ ||210 ||

रथिकाशिखाग्रचक्रं भ्रमचक्रारक (वि?) नियेजितं कार्यम् ।
यष्टिचुष्टयमसिंमस्तिर्यक् चक्रद्वयोपेतम् ॥211 ॥

rathikāśikhāgrachakraṁ bhramachakrāraka (vi?) niyejitaṁ
kāryam |
yaṣṭichuṣṭayamasiṁmastiryak chakradvayopetam ||211 ||

ऊर्ध्वं द्वितीयभूमेस्तृतीयभूमेरथान्तरे कुर्यात् ।
नियतं रथिकायष्टिभ्रमसंलग्नानि यन्त्राणि ॥212 ॥

ūrdhvaṁ dvitīyabhūmestṛtīyabhūmerathāntare kuryāt |
niyataṁ rathikāyaṣṭibhramasaṁlagnāni yantrāṇi ||212 ||

आसानाधारयष्टीनां रथिकाचक्रयोजितान् ।
अधः समान्तरान् कुर्याच्चतुरः परिवर्तकान् ॥213 ॥

āsānādhārayaṣṭīnāṁ rathikāchakrayojitān |
adhaḥ samāntarān kuryāchchaturaḥ parivartakān ||213 ||

त(द्व)द् द्वितीयभूमीदोलागर्भे समान्तरे यष्टी ।
लग्ने तथैकचक्रे याम्योत्तरचक्रयोर्न्यस्येत् ॥214 ॥

ta(dva)d dvitīyabhūmīdolāgarbhe samāntare yaṣṭī |
lagne tathaikachakre yāmyottarachakrayornyasyet ||214 ||

तद्वदधो भूकोणगरथिकाचूडाग्रचक्रसंसक्ताः ।
यष्टीस्ततश्चतस्रो द्विचक्रका इतरचक्रयोर्न्यस्येत् ॥215

tadvadadho bhūkoṇagarathikāchūḍāgrachakrasaṁsaktāḥ |
yaṣṭīstataśchatasro dvichakrakā itarachakrayornyasyet ||215

प्रान्तचक्रद्वये कोणरथिकाचक्रयोजिता ।
दोलागर्भगता यष्टिस्तिर्यक् कार्यापरापरा ॥216 ॥

prāntachakradvaye koṇarathikāchakrayojitā |
dolāgarbhagatā yaṣṭistiryak kāryāparāparā ||216 ||

पूर्वे भद्रे द्वारं कुर्यात् सोपानराजितमधस्तात् ॥
गर्भात् परिश्चमभागे निवेशयेद् देवतादोलाम् ॥217 ॥

pūrve bhadre dvāraṁ kuryāt sopānarājitamadhastāt ||
garbhāt pariśchamabhāge niveśayed devatādolām ||217 ||

अन्योन्यं चक्रभ्रममिच्छामुक्तिं विधानतः सम्यक् ।
ज्ञात्वा प्रयोजनीयं शीघ्रवहं मन्दवहनं वा ॥218 ॥

anyonyaṁ chakrabhramamichchhāmuktiṁ vidhānataḥ samyak |
jñātvā prayojanīyaṁ śīghravahaṁ mandavahanaṁ vā ||218 ||

एष समासेन यथा भ्रममार्गः कीर्तितः स्फुटोऽस्माभिः ।
अन्येष्वपि कर्तव्यः सम्यग् भ्रमहेतवे तद्वत् ॥ ।219 ॥

ēṣa samāsena yathā bhramamārgaḥ kīrtitaḥ

sphuṭo'smābhiḥ |
anyeṣvapi kartavyaḥ samyag bhramahetave tadvat || 1219 ||

स्तम्भादिद्रव्याणां विन्यासैः कल्पितं दृढैः श्लक्ष्णैः ।
सुश्लिष्टसन्धिबन्धं धृतं तथा दीर्घमुख्यघरैः ॥220॥

stambhādidravyāṇāṁ vinyāsaiḥ kalpitaṁ dṛdhaiḥ
ślakṣaṇaiḥ |
suśliṣṭasandhibandhaṁ dhṛtaṁ tathā
dīrghamukhyagharaiḥ || 220 ||

परिवारितमथ तिलकैः समन्ततः सिंहकर्णसंयुक्तम् ।
त्रिपुरं सम्यक् कुर्याद् विचित्ररूपं (स्व)कैश्चित्रैः ॥221॥

parivāritamatha tilakaiḥ samantataḥ
simhakarṇasaṁyuktam |
tripuraṁ samyak kuryād vichitrarūpaṁ
(sva)kaiśchitraiḥ || 221 ||

बुद्ध्या क्लृप्तैः पूर्वयन्त्रैश्च युक्तं
यन्त्राध्यायं वेत्ति यः सम्यगेतम् ।
प्राप्नोत्यर्थान् वांछितान् कीर्तियुक्तान्
स क्षमापालैरन्वह पूज्यते च ॥222॥ ।

buddhyā klṛptaiḥ pūrvayantraiścha yuktaṁ
yantrādhyāyaṁ vetti yaḥ samyagetam |
prāpnotyarthān vāṁchhitān kīrttiyuktān
sa kṣamāpālairanvaha pūjyate cha || 222 || |

एतद् द्वादशराजचक्रमखिलं क्षमापालचूडामणे-
र्दोःस्तम्भप्रतिबद्धवृत्ति परितो यस्येच्छया भ्रम्यति ।

ētad dvādaśarāājachakramakhilaṁ kṣamāpālachūḍāmaṇe-
rdoḥstambhapratibaddhavṛtti parito yasyechchhayā
bhramyati |

स श्रीमान् भुवनैकरामनृपतिर्देवो व्यधत्त द्रुतं
यन्त्राध्यायमिमं स्ववुद्धिरचितैर्यन्त्रप्रपंचैः सह ॥223॥

sa śrīmān bhuvanaikarāmanṛpatirdevo vyadhatta drutaṁ

yantrādhyāyamimaṁ svavuddhirachitairyantraprapaṁchaiḥ
saha ॥*223*॥

इति महाराजाधिराजश्रीभोजदेवविरचिते समराङ्गणसूत्रधारापरनाम्नि वास्तुशास्त्रे
यन्त्रविधानं नामैकत्रिंशोऽध्यायः ॥

iti mahārājādhirājaśrībhojadevavirachite samarāṅgaṇa-
sūtradhārāparanāmni vāstuśāstre yantravidhānaṁ nāmai-
katriṁśo' dhyāyaḥ ॥

References

Agarwal, D.P. (2000). *Ancient Metal Technology and Archaeology of South Asia: A Pan-Asian Perspective*, Aryan Books International, New Delhi.

Balasubramanian, B., (2002). *A New Study of Dhar Iron Pillar*, Indian Journal of History of Science, 37 (2), 115-151.

Kumar, Hemant. et al. (2015): *A Study and Brief Investigation on the Invisibility of Aircrafts (Vimanas)*, Intl. Journal of Engineering Research and Applications (Part 4) 5.1 (Jan. 2015).

Lafleur, Karl. et al. (2013): *Quadcopter Control in Three Dimensional Space using a Noninvasive Interface. J. Neural Engineering* (10.4).

Lahiri, N., (1995). *Indian Metal and Metal Related Artefacts as Cultural Signifiers: An Ethnographic Perspective, Word Archaeology*, 27 (1), 116–132.

Major, R.H ed. (1857): The travels of Niccolo Conti, *India in the Fifteenth Century*, Hakluyt Society, London, p. 27

Majumdar, R.C. (1978). *History of Ancient Bengal*, Calcutta.

Marco Polo, (1903). *The Book of Sir Marcopolo, the Venetian, concerning the kingdoms and marvels of the East.* Tr. and ed. by Henry Yule, London.

Ray, P.C., (1956). *A History of Hindu Chemistry*, Indian Chemical Society.

Ray, P.C. (1903). *History of Hindu Chemistry*, Vol 1. Calcutta.

Ray, P.C. (1909). *History of Hindu Chemistry*, Vol 2. Calcutta.

Samarāṅgaṇa Sūtradhāra, (1924 &1925): Ed. T. Ganapati Sastri, 2 Vols., Baroda

Shaw, R. L. (1985). *Fighter Combat: Tactics and Maneuvering.* Annapolis, Md: Naval Institute Press. ISBN 0-87021-059-9.

Śukranīti, (2005): B.K. Sarkar, J.P. Publishing House, Delhi.

Seal, B.N. (2008). *Hindu Chemistry*, Bharatiya Kala Prakashan, Delhi

Saligrama Krishna, Ramachandra Rao, (1985). *Encyclopaedia of Indian Medicine*, Volume 1, Popular Prakashan, Mumbai.

Sastri, K.A.N. (1966). *A History of South India.*

Sherby, Oleg D., and Jeffrey Wardsworth, (1985). *Damascus Steel, Scientific American*, 112–120, Feb. 1985.

Shruti K.R. et al. (2016). *Probable Technologies behind the Vimānas described in Rāmāyaṇa. Intl. Journal of Engineering Research and Applications.* Vol. 6, Issue 6 (Part 3). pp. 47–52.

Sigfried, J. de Laet, Joachim Herrmann, (1997). *History of Humanity: From the seventh century B.C. to the seventh century A.D.* United Nations.

Smith, C.S., (1982). *Damascus Steel, Science,* 216, 242–244.

Stein, (1907). *Ancient Khotan,* Vol.1 Oxford.

Travels of Marco Polo, The Venetian, (2003): John Masefield, New Delhi

Upadesh Manjari, (1998). *Poona Lectures* by Swami Dayanand Saraswati, Arsha Sahitya Prachar Trust, Delhi.

Vasco Da Gama: *A Journal of the First Voyage of Vasco Da Gama,* 1497-99: Tr. and ed. by E. G. Ravenstein, London. Hakluyt Society, p. 124 and n.4.

Wagner, D. B., (2007). *Chinese steelmaking techniques with a note on Indian wootz steel in China, Indian Journal of History of Science,* 42 (3), 289-318.

Vaidya, C.V. (1933). *Epic India,* Bombay.

Yuktikalpataru: Ed. Iswara Chandra Sastri.

Yule, H. (1914-17). *Cathay and the Way Thither.* Being a collection of Medieval Notices of China. Tr. and ed.

by Henry Yule. Vol.111. London. Hakluyt Society.
1914-17. pp. 66–67.

www.ingramcontent.com/pod-product-compliance
Lightning Source LLC
LaVergne TN
LVHW041450170726
843492LV00005B/1173
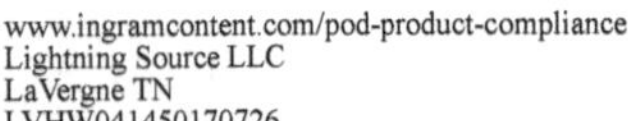